STUDENT STUDY GUIDE

Robert W. Christopherson
American River College - Emeritus

Charles E. Thomsen
American River College

Sixth Edition

Geosystems

An Introduction to Physical Geography

CHRISTOPHERSON

PEARSON

Prentice
Hall

Upper Saddle River, NJ 07458

Associate Editor: Amanda Griffith
Executive Editor: Dan Kaveney
Executive Managing Editor: Kathleen Schiaparelli
Assistant Managing Editor: Becca Richter
Production Editor: Gina M. Cheselka
Supplement Cover Manager: Paul Gourhan
Supplement Cover Designer: Joanne Alexandris
Manufacturing Buyer: Ilene Kahn
AV Art Director: Abigail Bass
AV Production Editor: Greg Dulles
Front Cover Photo: Grounded iceberg in Scoresby Sund, east-central Greenland–60-m (200-ft) tall "ice tower" by Bobbé Christopherson; Back Cover Photos: Arctic and Antarctic scenes by Bobbé Christopherson

© 2006 Pearson Education, Inc.
Pearson Prentice Hall
Pearson Education, Inc.
Upper Saddle River, NJ 07458

The author and publisher of this book have used their best efforts in preparing this book. These efforts include the development, research, and testing of the theories and programs to determine their effectiveness. The author and publisher make no warranty of any kind, expressed or implied, with regard to these programs or the documentation contained in this book. The author and publisher shall not be liable in any event for incidental or consequential damages in connection with, or arising out of, the furnishing, performance, or use of these programs.

Printed in the United States of America

10 9 8 7 6 5 4 3 2

ISBN 0-13-133092-6

Pearson Education Ltd., *London*
Pearson Education Australia Pty. Ltd., *Sydney*
Pearson Education Singapore, Pte. Ltd.
Pearson Education North Asia Ltd., *Hong Kong*
Pearson Education Canada, Inc., *Toronto*
Pearson Educación de Mexico, S.A. de C.V.
Pearson Education—Japan, *Tokyo*
Pearson Education Malaysia, Pte. Ltd.

CONTENTS

INTRODUCTION

WELCOME TO PHYSICAL GEOGRAPHY!

Thus begins the sixth edition of *Geosystems: An Introduction to Physical Geography* and this sixth edition of the Instructors Resource Manual. The success of the first five editions throughout the United States, Canada, and elsewhere is owed to the teachers and students who have embraced our approach to physical geography. The sixth edition is therefore dedicated to you the teachers and the students of this edition. *"To all the students and teachers of Earth, our home planet, and its sustainable future."*

During this edition of *Geosystems*, we celebrate the International Polar Year (IPY) from March 2007 to March 2009. This interdisciplinary research effort combines many scientific fields that are at the core of physical geography. Bringing together various disciplines is essential to understanding the ongoing dynamics of Earth's systems. The polar regions, which link to all systems across the globe, are experiencing profound change and much preparation is underway toward IPY research, exploration, and discovery (see: **http://dels.nas.edu/us-ipy/index.html**). You will see in *Geosystems* several high-latitude themes that place physical geography in a key place for spatial analysis of these global systems.

Physical geography provides essential information needed to understand and sustain our planetary journey in this new century. United Nations Secretary General Kofi Annan, speaking to the Association of American Geographers annual meeting, in 2001, offered this assessment:

> As you know only too well the signs of severe environmental distress are all around us. Unsustainable practices are woven deeply into the fabric of modern life. Land degradation threatens food security. Forest destruction threatens biodiversity. Water pollution threatens public health, and fierce competition for freshwater may well become a source of conflict and wars in the future. . . . the overwhelming majority of scientific experts have concluded that climate change is occurring, that humans are contributing, and that we cannot wait any longer to take action . . . environmental problems build up over time, and take an equally long time to remedy.

Physical geography is an excellent introduction and perhaps your only exposure to science and scientific methodology (see the Focus Study 1.1 "Scientific Method" in Chapter 1). Because of the nature of physical geography, a diverse and dynamic approach is at the heart of *Geosystems*.

Whatever you thought geography was going to be when you went through the catalog, schedule, and registration, remember that all of us are geographers at one time or another. Our mobility is such that we cover great distances every day.

Physical systems are expressed in complex patterns across Earth. Understanding both our lives and the planet's systems requires the tools of spatial analysis, which are at the core of geography. Physical geography is an important Earth system science, integrating many disciplines into a complete portrait of Earth, synthesizing many physical factors into a complete picture of Earth system operations.

A good example is the 1991 eruption of Mount Pinatubo in the Philippines. The global implications of this major event (one of the largest eruptions in the twentieth century) are woven through seven chapters of the book (see Figure 1.6 for a summary). Also, our update on global climate change and its related potential effects is part of the fabric in six chapters. These content threads, among many, weave together the variety of interesting and diverse topics crucial to a thorough understanding of physical geography.

The goal of physical geography is to explain the spatial dimension of Earth's natural systems—its energy, air, water, weather, climates, landforms, soils, plants, and life forms. For information about geography as a discipline and career possibilities contact the Association of American Geographers and ask for their pamphlets, "Careers in Geography" by Salvatore J. Natoli, "Geography: Today's Career for Tomorrow," and "Why Geography?" (see **http://www.aag.org**, or contact the National Council for Geographic Education, NCGE, at **http://www.ncge.org**).

Earth is a place of great physical and cultural diversity, yet people generally know little about it. Recent headlines have warned, "We Are Geographically Ignorant," or "Environmental Concerns Get Lost on the Map!" The 21st century is a time of critical questions and developments concerning human-environment themes, for many significant changes are already under way. Global climate change is at the forefront of issues confronting science, especially spatial sciences. Geography, as a spatial human–Earth science, is in a unique position among sciences to synthesize and integrate the great physical and cultural diversity facing us. Only through relevant education can an informed citizenry learn about the life-sustaining environment that surrounds and infuses our lives. To get started, complete the "Geography I.D." on pages 9 and 10 of this introduction.

By federal mandate, we now have an official "Geography Awareness Week" each November. Also, the National Geographic Society conducts a National Geography Bee beginning at the state level each April and ending with national finals held in Washington, D.C., in late May hosted by Alex Trebek from *Jeopardy!* The seventh National Geographic World Championship will be held in 2005 and broadcast on PBS. See Figure 21.10 for a photo of the last National Geographic Championship.

Geography is officially named as one of five key core curricula for national emphasis, along with history, mathematics, science, and English. As a discipline, geography underwent a strong resurgence throughout the 1990s.

Geography is a discipline in growing demand as humans place ever-increasing stress on the Earth–human relationship. Career opportunities involving geographic information systems (GIS) are flourishing in many fields. The demand for environmental analysis and assessment increases each year. We feel it important to keep these realities in mind as you progress through the book.

The purpose of this introduction is to give you some strategies for getting an "A" in this physical geography course and, ideally, to help you better perceive your relationship with Earth. Not bad for an initial goal statement! Also, a few hints are presented to make the overall learning process easier and possibly improve your success in all your classes. Juggling lecture and lab classes, study time, a job, family, and social activities is probably one of the toughest tasks in

life. And all are perhaps happening at a time when you may be dealing with the realization that these are the major leagues of life, the time and place for foundations to be built, and for careers and a meaningful future to be started. We stand in the first decade of a new millennium, with a valuable opportunity to reflect on methods and goals.

ON BEING A STUDENT

Teacher and student share common goals in education—thinking and learning. Teachers have taken on a lifetime association with the classroom, whereas the student passes through the classroom and on to a degree and a career. Most people think the goal of college is focused on grades, GPA, scholarship competition, and future job placement. Certainly and undeniably, these are important institutional goals and the immediate reward for doing well in classes. Ideally, however, the reward for being a student is thinking and learning. Thinking is the operation of the mind and a function of the brain, and learning represents a change in behavior.

Straight knowledge acquisition is at the simplest level of cognitive activity; applying that knowledge toward real action and behavior change is true learning. What if a student gets an "A" on a water conservation test and then goes home and wastes water? Knowledge acquisition occurred but not learning.

Thinking and learning, not grades, are the rewards for being a student. Grades are important and at the heart of institutional progress, but they are not the ultimate reward for being in school.

What about the teacher? How does this statement of setting idealistic goals apply to teaching about water conservation and then going home and wasting water? We can restate the axiom as follows: *thinking and learning, not pay, are the rewards for teaching.* Certainly pay is important, for teaching is a career and pay represents the teacher's livelihood. But pay is merely the institutional reason for participating. We, teacher and student, are linked in this critical educational pursuit of learning, positive behavioral change, and influencing the direction of society.

The reason for going into this in your study guide is to give you some insight into the educational philosophy of the author of the textbook that you are about to read.

GETTING ORGANIZED

A major aspect of your thinking and learning goals should be organization. Work out a detailed weekly schedule, by hour, for classes, study, work, your social life, sleeping, exercise, etc. Establish a routine around these activities, for time management is going to be a major organizational challenge. Find a comfortable study place on campus such as a specific carrel or table in the library.

For study at home, we recommend that students build a desk if they do not already have one. A plywood board (3/4" or 5/8") and cinder (masonry) blocks or chimney tiles, placed against a wall, and you're ready to go for under $30. Having a desk is extremely important to your college career, and quite superior to the bed, dining room table, floor, or couch in front of the television. Your desk is an excellent signal to others in your life that you are now serious about college and career building. Birthdays and gift days can be focused on this pursuit with equipment for classes: a desk pad, stapler, paper clips, pencils, pens, colored pencils, a calculator, protractor, metric ruler, manila folders, a collegiate dictionary, *Roget's Thesaurus* (an essential word finder), Strunk and White's *Elements of Style* (a guide to elementary grammar, word usage, and style), paper, binders, Post-It Notes, tab dividers, and, hopefully, a computer word-processing system.

Develop a system for organizing your binder, handouts, and notes; experiment with a system until you find one with which you are comfortable. We have seen bright students fall to poor grades by arriving in class with piles of disorganized papers. For each class, set up a method of coordinating reading notes with lecture notes. You may want to draw a dividing line on each sheet of binder paper about 7.5 cm (3 in.) in from the edge of the page. Reading notes can be placed on the left side and lecture notes on the right side of this line. Adjust this line for each class depending on the teacher's presentation.

STUDY METHODS AND THE TEXT

For your class notes, develop a personal style of outlining, hierarchical paraphrasing, succinct ways of recording lecture and reading notes. Be attentive to your teachers, for often they will tell you what items are important to know. The author has selected 350 figures from the text for overhead transparencies that your instructor may use, and they are on a PowerPoint presentation of the book. As these are used in lecture, your teacher will communicate the material to emphasize.

There is a range of opinion about the use of highlighting pens and the marking of sentence after sentence in each chapter. I'm not convinced that this works when compared with writing out succinct reading notes that correlate to your class notes. Working the content through your mind, translating it into your own words, and recording it in your handwriting (or keyboarding) impresses the material on your mind. The more conscious the process, the better the result. Underlining another person's words may not involve conscious mental activity on your part.

The author has taken care in writing *Geosystems* to consider the reader. Seven hundred **boldfaced** terms and concepts are presented with their definitions the first time they occur in the text and subsequently are used in the context of that definition. The glossary presents definitions that are keyed to the way the terms are used in the text. We recommend the use of flash cards with the terms or processes on one side and a sentence or two defining and using the terms on the other side. These cards can be used anywhere: at the mirror in the morning, at the bus stop, during meals, or at traffic lights.

Geosystems is organized in a way to assist the reader with a logical flow of topics. Subjects are presented in the sequence in which they occur in nature, or in a manner consistent with history and the flow of events. The author hopes this system organization will assist you with your reading. The entire text is organized around three orders of headings to give you a hierarchical nesting of subjects. These headings can form the basis of your outlining effort.

The author designed and wrote an important learning tool for each chapter. Note that a chapter begins with a list of "Key Learning Concepts." These tell you the concepts and activities that you should master upon reading the chapter. The operative words are highlighted in italics. Now turn to the end of a chapter and note the section titled "Summary and Review." Here you find definitions and a narrative, list of key terms with page numbers, and several review questions, all grouped under each key learning concept. This helps you coordinate your learning and your review. Every chapter concludes in a "Critical Thinking" section to stimulate additional questions and thought.

The textbook should provide you with many ideas for term paper and writing assignments that may arise in other classes.

Student Animation CD, by Robert Christopherson, is included in each copy of the text. This exciting learning tool contains 48 animations, satellite loops (including the 2004 Florida hurricanes), three exclusive photo galleries, a map reference library, and a searchable version of the text Glossary. Each animation features a self-test with pop-up reinforcement for correct answers. Text references provided throughout help you relate the CD to your use of the book.

METRIC SYSTEM

The text and all figures are in metric units and the International System of Units (SI) where appropriate. Canada uses the metric system exclusively, whereas English measurement equivalencies are presented for the United States, which is still in a transition period. A complete set of measurement conversions is conveniently presented in an easy-to-use arrangement in Appendix C of the text.

Student Lecture Notebook (ISBN 0-13-186353-3). All of the line art from the transparency set is reproduced in this full-color notebook. Students can now fully focus on the lecture and not be distracted by replicating drawings. Each page is three-hole punched for easy integration with other course materials.

THE INTERNET AND YOU

A wondrous tool is now in wide use—the Internet and the World Wide Web (WWW). *Geosystems* will expand its use and open new pathways into Earth systems sciences for you. And for the majority of you who regularly use this incredible resource, you already know what we mean.

The sixth edition of *Geosystems* presents about 200 URLs (Internet addresses) right in the text. If you find yourself intrigued by a subject you can go on the Net and find the latest information using these many links between the text and the Internet. We hope this innovation enlivens your teaching experience with the text.

The Internet and its World Wide Web is a resource that weaves threads of information from around the globe into a vast fabric. The fact that we now have Internet access into almost all the compartments aboard Spaceship Earth is clearly evident in *Geosystems*. Such ready availability of worldwide information allowed the author to illustrate content with a fascinating array of up-to-date examples and satellite images and to further verify accuracy.

Addresses on the Internet are known as URLs, which stands for Uniform Resource Locators—such as the one listed for the author's Prentice Hall "Geosystems Online Study Guide":
http://www.prenhall.com/christopherson.
Most home page addresses are obvious on an intuitive level. For example (after you enter *http://www.*): the Environmental Protection Agency is *epa.gov*; Natural Resources Defense Council is *nrdc.org*; Earth Observer System (NASA) is *eos.nasa.gov*; Netscape is *netscape.com*; and so on. ("Gov," "org," "edu," and "com" are obvious designations, among others now in use.) Most Internet software allows you to build up a "hot list" or set of "bookmarks" for the sites you visit frequently, so you have to enter or copy and paste their address only once.

A popular search engine is: **http://www.google.com**. To find information or to find a source for which you do not have an address, or simply to browse, enter the name of the organization or the subject matter as a search item and click. A listing of what you are looking for will usually appear in the first few hits. In this way any subject of interest can be researched.

A few examples. Let us assume that we want to find statistics for the predator–prey relationship between the moose and wolf populations on Isle Royale, an island in Lake Superior. First, enter the words Isle Royale wolves in "Google," one of the search engines. In short order several thousand items appear. The listing can be subdivided by subject area for easier browsing. One of the items on the list is "Ecological Studies of the Wolves and Moose of Isle Royale" reached at the URL:
http://www.isleroyalewolf.org/
This long address is quite typical and need not be entered because a click on the title (a linked item) enters it and accesses it for you.

Dr. Rolf Peterson's annual reports are summarized and the new data presented—moose count down, wolf count up for 2003–2004. The Web site even includes a brief QuickTime video of wolves chasing a moose. So, without typing anything and only using your mouse, this obscure fact is found quickly and more interesting leads are uncovered.

For instance, to better acquaint you with what it is like to live and work in the harsh climate at the South Pole, an account from people stationed there is carried on the Internet in *The New South Polar Times*, accessed at:
http://www.spotsylvania.k12.va.us/nspt/home.htm

The Naval Research Laboratory Monterey Satellite Meteorology home page accesses images and services related to global weather from a wide variety of satellite sources. Check it out:
http://www.nrlmry.navy.mil/sat_products.html

Our "Geosystems Online Study Guide" gives you online review, critical thinking exercises, tests that are graded online, opportunities to delve deeper into subjects out on the Net through our exclusive "Destinations" link feature that ties the chapter directly to numerous URLs, and follow-up answers to specific items in the text. Again, our URL (Net address) is:
http://www.prenhall.com/christopherson

To further help you, a booklet has been prepared titled *Science on the Internet: A Student's Guide* by

Andrew T. Stull, Prentice Hall (ISBN: 0-13-021308-X). This is a student's guide to the Internet and World Wide Web specific to geography. It is available free as a shrink-wrap with the text. If this did not come with your text please ask your teacher to contact the Prentice Hall sales representative about its availability and to receive your own copy. This guide also gives you a step-by-step procedure to set up your own home page.

The Internet changes rapidly and any list of URLs becomes dated quickly. The URLs in *Geosystems*, sixth edition, are accurate and up-to-date at press time; however, if you do come up with a "Not Found" oftentimes you can still find the source you are searching for by simplifying the URL. For example using our home page:

the *protocol* is http://
the *server* is www.prenhall.com
the *path* is /christopherson

Let's say that for some reason this URL does not work; if you delete items of the "path" you might score with just the protocol and server entries.

FORMAT FOR THIS STUDY GUIDE

Each chapter of this study guide matches the same numbered chapter in *Geosystems*. The format utilized for each study guide chapter is in the following sequence:

Outline Headings and Key Terms
URLs Presented in This Chapter
Key Learning Concepts for Chapter
Activities and Exercises (divided into ✷ **STEP** sections to assist you)
Sample Self-test (Answers appear at the end of this study guide.)

As you study each chapter, we recommend that you first preview it: look at the opening photograph, read the headings and the opening paragraph, leaf through and examine the figures and captions, sample topic sentences, and overview the chapter summary and review. Next, read the overview and learning objectives from this study guide. Use the Outline Headings and Key Terms list in this study guide as a checklist as you read the chapter.

As you read through the chapter, focus on the definition and application of each boldfaced term and check the glossary. Complete the activities and exercises in this study guide as you read the chapter

and attend class. End with the review questions at the end of the chapter. You may want to sketch answers in your reading notes for those review questions. Finish your work by taking the self-test in this study guide, then grade your test.

A TOUR OF *GEOSYSTEMS*

Geosystems is an introductory text in which relevant human-environmental themes are integrated with the principal core topics usually taught in a physical geography course. The text is conveniently organized into four parts that logically group chapters with related content. After completion of Chapter 1, "Essentials of Geography," these four parts may be studied in the integrated text sequence or in any order. Figure 1.8 illustrates the relation of these four parts and lists all the chapter titles.

The full-color figures, containing some 1,200 figure elements, many presented as compound, multimedia composites, evidence the author's commitment to high production values. *Geosystems*, sixth edition, contains more than 500 photographs (among these, 353 are by the author's wife and 98 were made by the author). We have more than 100 remote-sensing images, the most of any physical geography text. Hundreds of maps and artwork, and more than 50 tables, enhance the presentation and assist you in learning ideas and concepts.

Some of the text features include:

- Your study of physical geography begins with the front cover! The cover features an image of an "ice tower" in the Arctic. Climate change and its many impacts are hitting the high latitudes at a faster pace than elsewhere on Earth—approaching double the rate of change being experienced by the rest of the planet. To reflect global system linkages with these regions, *Geosystems*, sixth edition, weaves high-latitude themes through its chapters. Perhaps the most widespread climatic effect of global warming is rapid escalation of ice melt worldwide. Mount Kilimanjaro in Africa, portions of the South American Andes, and the Himalayas will very likely lose most of their glacial ice within the next two decades, affecting local water resources. Glacial ice continues its retreat in Alaska. NASA scientists determined that Greenland's ice sheet is thinning by about 1 m per year and numerous ice shelves in Antarctica are losing mass and even disintegrating. The "ice tower" iceberg cover photo symbolizes the majesty of Earth systems and the importance of science to analyze what is happening in the high latitudes and elsewhere. Remember, you are seeing about 14% of the iceberg in the photo. The other 86% is beneath the water, susceptible to the record warmth of ocean waters.

- Inside the front cover is a 2000 image of "The Living Earth" from Earth Imaging, Inc., Technical Director Erik Bruhwiler. The satellite-based computer illustration is created from hundreds of satellite images gathered by Advanced Very High Resolution Radiometer (AVHRR) sensors aboard several NOAA satellites. The text in Chapter 20 asks you to compare the global terrestrial biome map in Figure 20.3 with this illustration. "The Living Earth" features natural colors typical of local summer under cloudless conditions. Ocean-floor features are derived from other remotely sensed bathymetric data.

- New to this edition, mounted on the lower right-hand page of the inside-front cover is "Earth at Night," from the Operational Linescan System of the Defense Meteorological Satellite Program (DMSP), made in 2001. You can compare the Living Earth with the lights from human populations, on the same map projection.

- After Chapter 1, which presents the essentials of physical geography, including a discussion of geography, systems analysis, latitude and longitude, time, cartography, GIS, and remote sensing, a logical arrangement follows through the next 20 chapters, organized in four parts.

- *Geosystems* avoids the typical approach of tacked-on relevancy in isolated boxes or inserts, or placing them in the back of the book, or the omission of such material entirely. Care is taken to discuss applied topics with the core material where appropriate; in this way you can relate basic subject matter to the real world.

- Twenty-one "Focus Study" inserts, many with figures, highlight with greater depth key topics related to text material. The emphasis varies among the focus studies. Examples include the scientific method, stratospheric ozone losses, solar energy collection, wind power, the latest hurricane forecasting techniques, geothermal energy, floodplains, shoreline planning,

selenium concentration in soils, and the Great Lakes ecosystems. See the complete list of focus studies for the text at the end of this section.

- Fifty "News Report" features appear throughout the text. These cover items of timely or allied content: global positioning systems as a personal locator, harvesting drinking water from fog, water shortages in the Middle East, URLs for global climate change research, the disappearing Nile Delta, the ice core records from Greenland, alien species invasions, among many. News Report 3.1 in Chapter 3 details the highest sky dive through the stratosphere and troposphere. This feat provides an exciting way to learn about the layers in the lower atmosphere and the physical conditions experienced by a human exposed at such altitude. In addition, the author located a remote-triggered photo of Captain Kittinger as he jumped.

  New to this edition is the "High Latitude Connection" feature, linking chapter material to the polar regions. The world is turning its attention to the high latitudes to better understand present global climate change.

- The text and all figures use metric/English measurement equivalencies. A complete and expanded set of measurement conversions is conveniently presented in an easy to use arrangement in Appendix C of the textbook. For more information about metric conversion materials and programs contact the Metric Program Office, National Institute of Standards and Technology, U.S. Department of Commerce, Gaithersburg, MD 20899. A useful report was prepared by Barry N. Taylor, ed., *Interpretation of the SI for the United States and Metric Conversion Policy for Federal Agencies*, NIST Special Publication 814, NIST, Department of Commerce, October 1991. (See **http://www.nist.gov**.)

- You will notice the inclusion of the latest relevant scientific information throughout the text. This helps you see physical geography as an important Earth system science linked to other sciences in this era of expanded data gathering, computer modeling, and discovery. Geography is a dynamic science, not a static one!

- An up-to-date treatment of climate change, including global warming, is integrated throughout the text. You are shown how climate change specifically relates to many aspects of physical geography. Note that the topic does not appear as one of the focus studies; rather, it is treated within the text. There is a scientific consensus relative to global warming. All the reports from the Intergovernmental Panel on Climate Change through 2001 provide an important resource for these sections. A specific section on global climate change is featured in Chapter 10.

- Currency has always been a feature, and this edition is no exception. Here are a few examples to illustrate this point: the latest air pollution information, tornado and hurricane data through 2004, global temperature graph updated through 2003, the latest in weather technology (including ASOS, AWIPS), earthquake and volcano eruption data through 2004, South Cascade Glacier mass balance data through 2002, Great Lakes levels through 2001, the agenda of the 2002 Earth Summit (as it relates to physical geography), results of the November 2001 Marrakech, Morocco, Climate Summit, El Niño/La Niña status and 2004 forecast, new sulfur deposition maps through 1999, and summaries of the Intergovernmental Panel on Climate Change 2001 reports.

- In key chapters (climate, arid landscapes, soils, biomes), large, integrative tables are presented to help synthesize content. Following each of these tables, if you choose, the rest of the chapter can be studied for deeper analysis of the subject. In fact, Table 20.1 in Chapter 20 can be referenced throughout the term, as the total Earth system is organized within the various terrestrial biomes.

- Compound figures appear in every chapter. These integrate art, photos, and maps to demonstrate text concepts, as developed by this author from many years of using a multimedia approach in classroom teaching.

- The Glossary in *Geosystems* is prepared with text-related definitions and explanations for 650 terms and concepts. Over 1800 terms and concepts appear in the relational index. Tab it with a Post-It Note.

- The Career Link features are updated and the author visited several of the interviewees for updating and photos. A new Career Link in Chapter 5 features Dr. Louwrens Hacquebord, Professor of Arctic and Antarctic Studies at the University of Groningen in the Netherlands. This new Career Link weaves together the High Latitude Connection, applied geography (including GIS), and accelerated climate change in the Arctic, as the author takes you through an excavation of a whaling site in southwestern Spitsbergen.

- Appendix A contains a discussion of maps used in this text and topographic maps.

- Appendix B presents the Köppen climate classification system, including Figure B.1, which ties the Köppen climate classification system to the climate classification system in Figure 10.5.

- *Geosystems* ends with a capstone chapter, "Earth and the Human Denominator," which is unique in physical geography. *Geosystems* attempts to look holistically at Earth and the world in this concluding treatment. The chapter begins with a discussion of the human population count, both in population numbers and in the impact per person. This chapter can open up class discussions concerning the impact of humans on Earth systems, as well as the important role played by geographic analysis and physical geography in particular. Mention is made of the second Earth Summit held in 2002 in Johannesburg, South Africa, with an agenda including climate change, freshwater, and the five Rio Conventions, among other topics.

Please read the *Geosystems* Preface for a complete list of text features and support resources.

MAP ASSIGNMENT

In Chapter 5 of the text the following suggestion appears:

With each map, begin by finding your own city or town and noting the temperatures indicated by the isotherms. Record the information from these maps in your notebook. As you work through the different maps throughout this text, note atmospheric pressure and winds, precipitation, climate, landforms, soil orders, vegetation, and terrestrial biomes. By the end of the course you will have recorded a complete profile of your regional environment.

We recommend that you follow through on this suggestion and maybe even establish a special section in your notebook. You might want to expand your entries to include your state or province and the general region where you live.

The "Geography I.D." exercise on the next two pages can be drawn out over the entire term and completed as topics are covered in the text. The map and the I.D. effort will give you a geographic sense of your location, place, and regional setting.

Our contact information:

Robert W. Christopherson
P.O. Box 128
Lincoln, CA 95648
E-mail: bobobbe@aol.com
http://www.prenhall.com/christopherson

Charles E. Thomsen
American River College
4700 College Oak Drive
Sacramento, CA 95841
E-mail: thomsec@arc.losrios.edu
http://ic.arc.losrios.edu/~thomsec

THE "GEOGRAPHY I.D." ASSIGNMENT FOR STUDENTS:

GEOGRAPHY I.D.

To begin: complete your personal Geography I.D. Use the maps in your text, an atlas, college catalog, and additional library materials if needed.

Your instructor will help you find additional source materials for data pertaining to the campus.

On each map find the information requested noting the January and July temperatures indicated by the isotherms (the small scale of these maps will permit only a general determination), January and July pressures indicated by isobars, annual precipitation indicated by isohyets, climatic region, landform class, soil order, and ideal terrestrial biome.

Record the information from these maps in the spaces provided. The completed page will give you a relevant geographic profile of your immediate environment. As you progress through your physical geography class the full meaning of these descriptions will unfold. This page might be one you will want to keep for future reference.

[Derived from *Applied Physical Geography: Geosystems in the Laboratory*, sixth edition, by Robert W. Christopherson and Charles E. Thomsen, © 2006 Prentice Hall, Inc., Pearson Education]

GEOGRAPHY I.D.

NAME: _____ LAB SECTION: _____

HOME TOWN: _____ LATITUDE: _____ LONGITUDE: _____

COLLEGE/UNIVERSITY: _____

CITY/TOWN: _____ COUNTY (PARISH): _____

    Standard Time Zone (for College Location): _____

    Latitude: _____ Longitude: _____

    Elevation (include location of measurement on campus): _____

    Place (tangible and intangible aspects that make this place unique): _____

    _____

    Region (aspects of unity shared with the area; cultural, historical, economic, environmental):

    _____

    _____

    Population: Metropolitan Statistical Area (CMSA, PMSA, if applicable): _____

    Environmental Data: (Information sources used): _____

    January Avg. Temperature: _____ July Avg. Temperature: _____

    January Avg. Pressure (mb): _____ July Avg. Pressure: _____

    Average Annual Precipitation (cm/in.): _____

    Avg. Ann. Potential Evapotranspiration (if available; cm/in.): _____

    Climate Region (Köppen symbol and name description): _____

    _____

    Main climatic influences (air mass source regions, air pressure, offshore ocean temperatures, etc.):

    _____

    _____

    Topographic Region or Structural Region (inc. rock type, loess units, etc.): _____

    _____

    Dominant Regional Soil Order: _____

    Biome (terrestrial ecosystems description; ideal and present land use): _____

    _____

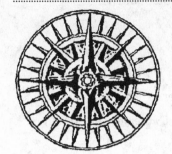

1

ESSENTIALS OF GEOGRAPHY

Chapter 1 features a quote from the past president of the American Association for the Advancement of Science that demonstrates the need for the science of physical geography. Read this to begin your study of *Geosystems*.

> . . . during the last few decades, humans have emerged as a new force of nature. We are modifying physical, chemical, and biological systems in new ways, at faster rates, and over larger *spatial* scales than ever recorded on Earth. Humans have unwittingly embarked upon a grand experiment with our planet. The outcome of this experiment is unknown, but has profound implications for all of life on Earth.*

"Essentials of Geography" contains the basic tools needed to study the content of physical geography: spatial analysis, systems, models, the physical planet, geographic coordinates of latitude and longitude, time, cartography and map making, remote sensing, and geographic information systems (GIS). With the completion of this chapter you have the "tool kit" necessary to work through the four parts and twenty chapters ahead.

As you study, note that the figures, text, and this study guide are carefully integrated to help you learn physical geography. I wrote the figure captions as an important part of this process. You might want to begin each chapter by turning pages and examining the photographs, illustrations, maps, and their captions.

OUTLINE HEADINGS AND KEY TERMS

The first-, second-, and third-order headings that divide Chapter 1 serve as an outline for your notes

---

* Jane Lubchenco, Presidential Address, American Association for the Advancement of Science, February 15, 1997.

and studies. The key terms and concepts that appear **boldface** in the text are listed here under their appropriate heading in ***bold italics***. All these highlighted terms appear in the text glossary. Note the check-off box (❏) so you can mark your progress as you master each concept. These terms should be in your reading notes or used to prepare note cards.

The outline headings and terms for Chapter 1:

❏ *Earth systems science*

The Science of Geography

❏ *geography*
❏ *spatial*
❏ *location*
❏ *region*
❏ *human–Earth relationships*
❏ *movement*
❏ *place*

Geographic Analysis

❏ *spatial analysis*
❏ *process*
❏ *physical geography*
❏ *scientific method*

The Geographic Continuum

Earth Systems Concepts

Systems Theory

❏ *system*

Open Systems

❏ *open system*

Closed Systems

❏ *closed system*

System Example

Focus Study 1.2: The Timely Search for Longitude

Career Link 1.1: Thomas D. Jones, Ph.D., Astronaut, Earth Observer, and Geographer

High Latitude Connection 1.1: Meltponds as Positive Feedback

URLs listed in Chapter 1

International Polar Year:
http://dels.nas.edu/us-ipy/index.html

Geography organizations:
http://www.amergeog.org/
http://www.aag.org/
http://www.ncge.org/
http://www.nationalgeographic.com/
http://venus.uwindsor.ca/cag/cagindex.html
http://www.rcgs.org/
http://www.iag.org.au/
http://www.agta.asn.au/
http://www.egea.geog.uu.nl/
http://www.rgs.org/
*http://www.geoggeol.fau.edu/prof_org/
    GeographyPO.html*

Time and UTC:
*http://www.gmt2000.co.uk/meridian/place/
    plco0a1.htm*
http://aa.usno.navy.mil/faq/docs/world_tzones.html
http://www.bipm.fr/
http://www.nrc.ca/
http://www.bipm.fr/en/bipm/
http://nist.time.gov/
http://webexhibits.org/daylightsaving/

Maps:
http://nationalmap.usgs.gov/

Satellites and remote sensing:
http://www.gsfc.nasa.gov/indepth/earth_esm.html
http://www.gsfc.nasa.gov/
http://envisat.esa.int/
http://geo.arc.nasa.gov/sge/landsat/landsat.html
http://www.noaa.gov/
http://terra.nasa.gov/
http://www.goes.noaa.gov/
http://rsd.gsfc.nasa.gov/goes/
http://www.ghcc.msfc.nasa.gov/GOES/
http://www.vtt.fi/tte/research/tte1/tte14/virtual/
http://spot4.cnes.fr/waiting.htm
http://www.spaceimaging.com/
http://edc.usgs.gov/

GIS and GPS:
http://www.geo.ed.ac.uk/home/giswww.html
http://erg.usgs.gov/isb/pubs/gis_poster/
http://geogratis.cgdi.gc.ca/CLI/frames.html
http://nationalmap.usgs.gov/
*http://www.colorado.edu/geography/gcraft/notes/gps/
    gps_f.html*
http://www.geo.ed.ac.uk/home/giswww.html
http://www.gsfc.nasa.gov/topstory/20020828phap.html
*http://www.cdc.gov/ncidod/dvbid/westnile
        http://www.arc.losrios.edu/~earthsci/GIS/
        GIS.html*
http://www.ncgia.ucsb.edu/
http://www.ncgia.maine.edu/
http://www.geog.buffalo.edu/ncgia/
*http://www.spaceflight.nasa.gov/gallery/images/shuttle/
    sts-98/ndxpage1.html*

KEY LEARNING CONCEPTS FOR CHAPTER 1

The following key learning concepts help guide your reading and comprehension efforts. The operative word is in *italics*. Use these carefully to guide your reading of the chapter and note that STEP 1 asks you to work with these concepts. These same learning concepts are used in organizing the summary and review section at the end of the chapter—grouping together definitions, a list of key terms, and review questions.

After reading the chapter and using this study guide, you should be able to:

- *Define* geography and physical geography in particular.
- *Describe* systems analysis, open and closed systems, feedback information, and system operations, and *relate* these concepts to Earth systems.
- *Explain* Earth's reference grid: latitude, longitude, and latitudinal geographic zones and time.
- *Define* cartography and mapping basics: map scale and map projections.
- *Describe* remote sensing and *explain* geographic information system (GIS) methodology as a tool used in geographic analysis.

✳ STEP 1: Critical Thinking Process

Using your interest and learning, and the following questions as guidelines <u>only</u>, briefly discuss your experience with this chapter. In examining your learning you need not go through each of these questions in detail, simply provide an overview of your critical thinking process as it relates to some aspect of this chapter.

- What did you know about the learning concept before you began?
- Which information sources did you use in your learning (text, class, other)?
- Were you able to complete the action stated in the learning concept? What did you learn?
- Are there any aspects of the concept about which you want to know more?

Critical Thinking and Chapter 1: _____

_____

_____

_____

_____

✳ STEP 2: Five Themes of Geographic Science

Using Figure 1.1, analyze and describe why you think the author selected the five photographs to represent each of the geographic themes of geographic science.

1. Location:_____

_____

_____

2. Place: _____

_____

_____

3. Movement:_____

_____

_____

4. Regions: _____

_____

_____

5. Human–Earth relationships: _____

_____

_____

✳ STEP 3: Critical Environmental Concerns and Physical Geography

The past president of the AAAS said, " . . . humans have emerged as a new force of nature." In March 2001, United Nations Secretary General Kofi Annan, and recipient of the 2001 Nobel Peace Prize, spoke to the Association of American Geographers annual meeting. In part, he said,

> As you know only too well the signs of severe environmental
> distress are all around us. Unsustainable practices are woven
> deeply into the fabric of modern life.

List, in your own words, several of the critical environmental concerns mentioned in the text—leaf through the book, check the index, check Chapter 21, "Earth and the Human Denominator." Following a possible class discussion or talking with others, add any two additional concerns that you think should be considered in physical geography. State your reasons for me to add these items to a future edition of *Geosystems*.

1. _____

_____

2. _____

_____

3. _____

_____

4. _____

_____

5. _____

_____

6. _____

_____

7. Additional concern: _____

_____

_____

8. Additional concern: _____

_____

_____

✳ STEP 4: Systems Theory

Inside the box provided on the next page, <u>sketch</u> a simple open system as in Figures 1.3 and 1.4. <u>Indicate</u> flows of energy and matter into and out of the system and within the system. Identify the components of a system. Your sketches may be schematic line drawings or an actual depiction of a real system as in Figure 1.3 (automobile) or Figure 1.4 (leaf). Below your sketch, <u>describe</u> a representative example.

1. **Open** system

Example: _____

_____

_____

_____

2. After reading High Latitude Connection 1.1 describe the feedback links in the meltpond–climate system.

_____

_____

_____

3. Compare Earth's climate to the diagrams of steady-state and dynamic equilibria (Figure 1.5, p. 10). Which diagram does our climate most closely resemble? Are there any indications that we may be switching to a different type of equilibrium? _____

_____

_____

✳ STEP 5: A Spherical Planet and Location

1. Earth's equatorial diameter is _____ km (_____ mi).

2. Earth's polar diameter is _____ km (_____ mi). Explain why these diameters are different: _____

_____.

In this modern era, Earth's shape is described as a _____.

3. What is your present latitude? _____ ; longitude? _____

What was your source for this information (Internet, library, instructor, atlas, encyclopedia, topographic map, local agency or organization, benchmark on campus, GPS unit)? Describe. _____

_____

_____.

Record these two values on your "Geography I.D." sheet in the Study Guide introduction.

4. Properly label the following illustrations from Figures 1.11 and 1.14 in the spaces provided.

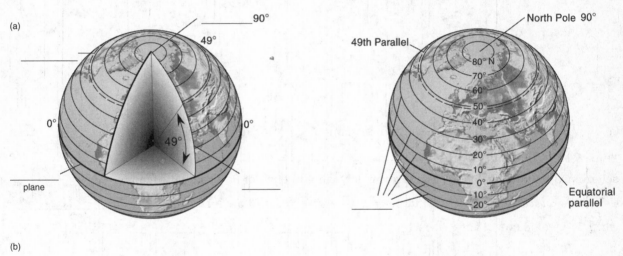

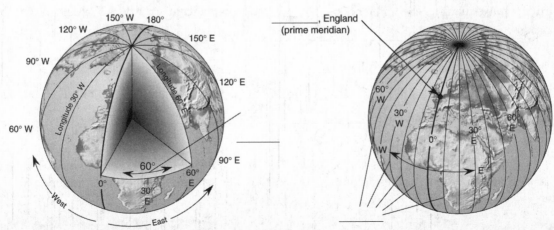

5. Relative to Eratosthenes and his determinations of Earth's circumference in 247 B.C., complete the labeling of the following figure from Figure 1.10. (The lines are leader lines pointing to places in the illustration. Use the flat line at the end of the leader line for your label.)

Eratosthenes' Calculation (Figure 1.10):

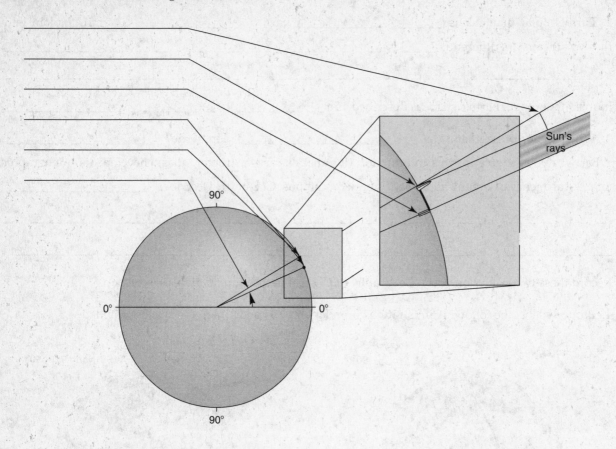

(a) How many stadia separated Syene and Alexandria according to Eratosthenes? _____. Each stadia represented approximately _____ m (_____ft). He determined that this distance was approximately what fraction of Earth's total circumference? _____. Using this information, Earth's circumference is represented by how many stadia? _____; which is equal to how many meters? _____ m (_____ft).

Given Earth's actual equatorial circumference of _____m, what percent error does Eratosthenes' measurement represent? _____%.

(b) What is the science of Earth's shape called? _____

(c) Columbus did not discover Earth's sphericity. According to the text what ancient scholar is credited with first deducing that Earth was a sphere? _____

6. Table 1.2 presents the linear distance for degrees of latitude and degrees of longitude at various latitudes. A small diagram of these values accompanies the table. Please complete the labels on the following illustration.

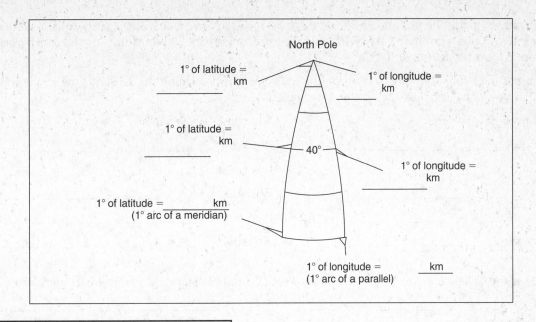

✳ STEP 6: Standard Time Zones

Because Earth rotates 360° every 24 hours, or 15° per hour (360° ÷ 24 = 15°) along any parallel, a time zone of one hour is established for each 15° of longitude. Each time zone theoretically covers 7.5° on either side of a *controlling meridian* (0°, 15°, 30°, 45°, 60°, 75°, 90°, 105°, 120°, etc.) and represents one hour. Figure 1.17 is printed below to help you, although the color version in the text is best to use.

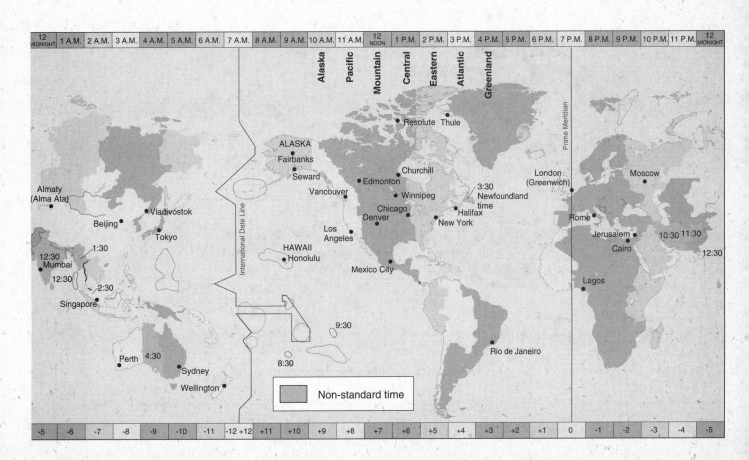

1. The numbers along the bottom margin of the map indicate how many hours each zone is earlier (plus sign) or later (minus sign) than Greenwich and the prime meridian. Standard time is broadcast worldwide as *Coordinated Universal Time*, or *UTC*.

(a) From the map of global time zones (Figure 1.17) can you determine the present time in the following cities? (For your local time use the time at this moment.)

Your present local time? _____ Your time zone's name: _____

Moscow? _____ Denver? _____

London? _____ Los Angeles? _____

Halifax? _____ Fairbanks? _____

Chicago? _____ Honolulu? _____

Winnipeg? _____ Singapore? _____

(b) What is the distance in km (mi) between your college and the standard (controlling) meridian for your time zone? _____.

2. What is UTC? Briefly describe it and how it is determined. _____.

_____

_____

3. According to Figure 1.18 in the text and your understanding of the concepts presented, label or describe all the components in this illustration.

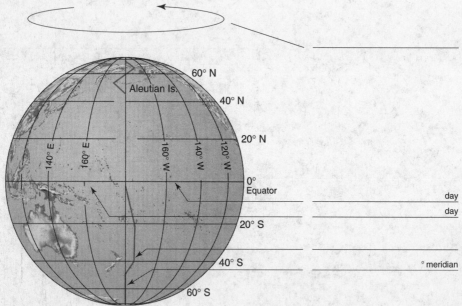

✳ STEP 7: Map Projections

1. What is a map projection? List some ways map projections are different from the globe.

_____

_____

_____.

2. What are the four classes of map projections (Figure 1.22)?

_____

_____

3. List the *representative fraction* for the following: a small-scale map: _____ ;

and a large-scale map: _____

✳ STEP 8: Remote Sensing, GIS, and GPS

1. Using Figure 1.24 and the text, define remote sensing: _____.

_____

_____

2. Relative to remote sensing, distinguish between a "photograph" and an "image."

(a) Photograph: _____;

_____

(b) Image: _____

_____

3. There are 105 remote-sensing images in *Geosystems*. Distinguish between active and passive remote sensing systems by leafing through this book and identifying three examples of each type (give figure number and title).

Active remote sensing: **(a)** _____,

(b) _____,

(c) _____.

Passive remote sensing: **(a)** _____,

(b) _____,

(c) _____.

4. What is GIS? Briefly describe. _____

_____

_____

5. How might a GPS unit and remote sensing imagery be used in the preparation of a GIS model? Briefly describe. _____

_____

_____

There are many Internet addresses (URLs) listed in this chapter of the *Geosystems* textbook. Go to any two URLs, or "Destination" links on the *Geosystems* Home Page, and briefly describe what you find.

1. _____ : _____.

_____

2. _____ : _____.

_____

SAMPLE SELF-TEST

1. The word *spatial* refers to

    a. the how and why questions rather than the where question
    b. items that relate specifically to society
    c. the nature and character of physical space
    d. eras of time, important to history

2. Relative to the fundamental themes of geography authored by the Association of American Geographers and the National Council for Geographic Education, latitude and longitude refer to

    a. location
    b. place
    c. movement
    d. regions

3. The three most costly natural disasters in history in terms of property damage—Hurricane Andrew, the Midwest floods, and the Northridge (Reseda) quake—fall within which theme?

    a. place
    b. human–Earth relationships
    c. movement
    d. regions

4. The scientific method is described by which of the following?

    a. the application of common sense
    b. the development of hypotheses for testing and prediction
    c. the formation of theories encompassing broad general principles
    d. all of these are correct

5. Earth's environment, operating as a system, is characterized by which one of the following?

    a. an open system in terms of both energy and matter (air, water, soil)
    b. a closed system in terms of energy
    c. entirely abiotic
    d. a closed system in terms of matter (air, water, soil)

6. According to the text, the three *inorganic* Earth realms are the

    a. thermosphere, lithosphere, heterosphere
    b. hydrosphere, lithosphere, atmosphere
    c. atmosphere, geoid, homosphere
    d. stratosphere, magnetosphere, troposphere

7. Earth's circumference was first calculated by

   a. Columbus
   b. modern satellite measurements
   c. Eratosthenes, the librarian at Alexandria
   d. Sir Isaac Newton

8. The basis of time is the fact that Earth

   a. rotates from east to west, or westward
   b. GMT is measured in Washington, D.C.
   c. moves around the Sun taking 365.25 days a year
   d. rotates on its axis in 24 hours, or 15° per hour

9. UTC refers to

   a. Universal Time Conference
   b. Coordinated Universal Time
   c. United States time
   d. a system of local time using phases of the moon

10. Which of the following possesses *all* of Earth's surface properties of area, shape, direction, proximity, and distance?

    a. Goode's homolosine projection
    b. Mercator projection
    c. Alber's equal-area conic projection
    d. a globe

11. Geography is defined by its spatial approach rather than by a specific body of knowledge or content.

    a. true
    b. false

12. The tangible and intangible aspects of a region specifically refer to the geographic theme of *location*.

    a. true
    b. false

13. Negative feedback, most common in nature, tends to discourage response in a system, promoting system self-regulation.

    a. true
    b. false

14. The precise determination of longitude at sea was impossible until as late as 1760 A.D.

    a. true
    b. false

15. That part of geography that embodies mapmaking is called geodesy.

    a. true
    b. false

16. The Eastern Hemisphere side of the International Date Line (180th meridian) is always one day *ahead* of the Western Hemisphere side.

    a. true
    b. false

17. According to Appendix A, only the Goode's homolosine map projection possesses *both* the qualities of equal area and true shape.

    a. true
    b. false

18. Geography is a _____ science.

19. In terms of energy Earth is an _____ system, whereas in terms of matter Earth is a _____ system. A modern landfill (dump) is indicative of the fact that Earth is a _____ material system.

20. From anywhere in the Northern Hemisphere, you can determine your latitude by sighting on _____ at night. In the Southern Hemisphere the _____ constellation is used to locate the celestial pole.

21. Coordinated Universal Time (UTC) is now measured by _____ clocks. The newest United States clock is named the _____, operated by Time and Frequency Services of the N.I.S.T., which stands for: _____.

22. A great circle is _____; a small circle is _____.

23. The four classes of map projections are:

    (a) _____   (c) _____

    (b) _____   (d) _____

24. Which of the following is the most widely used map prepared by the USGS?
    a. topographic map
    b. a Robinson projection
    c. a land resources map
    d. a soils map

25. At least three methods for expressing map scale are:

    a. _____

    b. _____

    c. _____

26. In the environment about us _____ is an entity that assumes a physical shape and occupies space; whereas _____ is a capacity to do work.

27. Which map projection from the book is used for the following?
    a. Figure 1.13: _____

    b. Figure 1.17: _____

    c. Figure 1.22a: _____

    d. Appendix B, Figure B.1: _____

PART ONE:
The Energy–Atmosphere System

OVERVIEW—PART ONE

Part One exemplifies the systems organization of the text: Part One of *Geosystems* begins with the Sun and Solar System (see the systems diagram in Figure 1.7). Solar energy passes across space to Earth's atmosphere, where it is distributed unevenly as a result of Earth's curvature. Physical factors produce seasonal variations of daylength and Sun altitude as the subsolar point migrates between the tropics (Chapter 2).

The atmosphere acts as a membrane, editing out harmful wavelengths of electromagnetic energy and protecting Earth from the solar wind and most debris from space. The lower atmosphere is being dramatically altered by human activities (Chapter 3). Insolation passes through the atmosphere to Earth's surface, where patterns of surface energy budgets are produced (Chapter 4). These flow processes produce world temperature patterns (Chapter 5), and general and local atmospheric and oceanic circulations are created (Chapter 6). These five chapters portray the Earth's energy–atmosphere system.

Name: _____   Class Section: _____

Date: _____Score/Grade: _____

2

SOLAR ENERGY TO EARTH AND THE SEASONS

Our planet and our lives are powered by radiant energy from the star closest to Earth—the Sun. In this chapter we follow the Sun's output to Earth and the top of the atmosphere. The uneven distribution of insolation sets the stage for all the motions and flow systems we study in later chapters.

Incoming solar energy that arrives at the top of Earth's atmosphere establishes the pattern of energy input that drives Earth's physical systems and that daily influences our lives. This solar energy input to the atmosphere, plus Earth's tilt and rotation, produce daily, seasonal, and annual patterns of changing daylength and Sun angle. The Sun is the ultimate energy source for most life processes in our biosphere.

Think for a moment of the annual pace of your own life, your wardrobe, gardens, and lifestyle

activities—all reflect shifting seasonal energy patterns. Seasonality—the periodic rhythm of warmth and cold, dawn and daylight, twilight and night—affects all of our lives and has fascinated humans for centuries.

OUTLINE HEADINGS AND KEY TERMS

The first-, second-, and third-order headings that divide Chapter 2 serve as an outline for your notes and studies. The key terms and concepts that appear **boldface** in the text are listed here under their appropriate heading in ***bold italics***. All these highlighted terms appear in the text glossary. Note the check-off box (❏) so you can mark your progress as you master each concept. These terms should be in your reading notes or used to prepare note cards. The ⊙ icon indicates that there is an accompanying animation on the Student CD.

The outline headings and terms for Chapter 2:

The Solar System, Sun, and Earth

- ❏ *Milky Way Galaxy*

⊙ **Nebular Hypothesis**

Solar System Formation and Structure

- ❏ *gravity*
- ❏ *planetesimal hypothesis*

Dimensions and Distances

- ❏ *speed of light*

Earth's Orbit

- ❏ *perihelion*
- ❏ *aphelion*
- ❏ *plane of the ecliptic*

Solar Energy: From Sun to Earth

⊙ **Electromagnetic Spectrum and Plants**

- ❏ *fusion*

Solar Activity and Solar Wind

- ❏ *solar wind*
- ❏ *sunspots*

Solar Wind Effects

- ❏ *magnetosphere*
- ❏ *auroras*

Weather Effects

Electromagnetic Spectrum of Radiant Energy

- ❏ *electromagnetic spectrum*
- ❏ *wavelength*

Incoming Energy at the Top of the Atmosphere

- ❏ *thermopause*
- ❏ *insolation*

Solar Constant

- ❏ *solar constant*

Uneven Distribution of Insolation

- ❏ *subsolar point*

Global Net Radiation

The Seasons

⊙ **Earth–Sun Relations, Seasons**

Seasonality

- ❏ *altitude*
- ❏ *declination*
- ❏ *daylength*

Reasons for Seasons

Revolution

- ❏ *revolution*

Rotation

- ❏ *rotation*
- ❏ *axis*
- ❏ *circle of illumination*

Tilt of Earth's Axis

- ❏ *axial tilt*

Axial Parallelism

- ❏ *axial parallelism*

Sphericity

Annual March of the Seasons

- ❏ *sunrise*
- ❏ *sunset*
- ❏ *inter solstice*
- ❏ *December solstice*
- ❏ *ernal equinox*
- ❏ *arch equinox*
- ❏ *ummer solstice*
- ❏ *June solstice*
- ❏ *utumnal equinox*
- ❏ *September equinox*

Dawn and Twilight

Seasonal Observations

SUMMARY AND REVIEW

News Reports

News Report 2.1: The Nature of Order Is Chaos
News Report 2.2: Monitoring Earth Radiation Budget

URLs listed in Chapter 2

Solar systems simulator:
http://space.jpl.nasa.gov/

Sunspot cycle and auroral activity:
*http://wwwssl.msfc.nasa.gov/ssL/pad/solar/
    sunspots.htm*
http://www.sel.noaa.gov/pmap/
http://www.gi.alaska.edu
http://northernlightsnome.homestead.com/

International Earth Rotation Service:
http://www.iers.org/

Sunrise, sunset calculator:
http://www.srrb.noaa.gov/highlights/sunrise/sunrise.html

URLs of Interest

Hubble telescope and astronomy info:
http://sohowww.nascom.nasa.gov/gallery/
http://www.seds.org/hst/
http://www.solarviews.com/

KEY LEARNING CONCEPTS FOR CHAPTER 2

The following key learning concepts help guide your reading and comprehension efforts. The operative word is in *italics*. Use these carefully to guide your reading of the chapter and note that STEP 1 asks you to work with these concepts. These same learning concepts are used in organizing the summary and review section at the end of the chapter—grouping together definitions, a list of key terms, and review questions.
    After reading the chapter and using this study guide, you should be able to:

- *Distinguish* among galaxies, stars, and planets, and *locate* Earth.
- *Overview* the origin, formation, and development of Earth and *construct* Earth's annual orbit about the Sun.
- *Describe* the Sun's operation and *explain* the characteristics of the solar wind and the electromagnetic spectrum of radiant energy.
- *Portray* the intercepted solar energy and its uneven distribution at the top of the atmosphere.
- *Define* solar altitude, solar declination, and daylength and *describe* the annual variability of each—Earth's seasonality.

✳ STEP 1: Critical Thinking Process

Using your interest and learning, and the following questions as guidelines <u>only</u>, briefly discuss your experience with this chapter. In examining your learning you need not go through each of these questions in detail, simply provide an overview of your critical thinking process as it relates to some aspect of this chapter.

- What did you know about the learning concept before you began?
- Which information sources did you use in your learning (text, class, other)?
- Were you able to complete the action stated in the learning concept? What did you learn?
- Are there any aspects of the concept about which you want to know more?

Critical Thinking and Chapter 2: _____

_____

_____

_____

_____

✳ STEP 2: Some Initial Concepts

1. Briefly summarize present scientific thinking on how the Sun, Earth, and other planets formed.

_____

_____

_____

_____

2. What is going on in Figure 2.3? Why not deploy this unrolled foil on Earth and avoid all the trouble of going to the Moon?

_____

_____

_____

3. Briefly describe several of the effects created by the solar wind in the atmosphere.

_____

_____

_____

4. How does the Sun generate so much energy? What process is occurring deep within the Sun?

_____

_____

_____

✳ STEP 3: The Solar System; Earth's Orbit

In the space provided, make a simple <u>sketch</u> of Earth's orbit about the Sun ("S"); <u>identify</u> **perihelion** and **aphelion** locations, calendar dates, and orbital distances at the two extreme points in the orbit.

✳ STEP 4: Electromagnetic Spectrum; Insolation

1. A portion of the electromagnetic spectrum of radiant energy is emitted by the Sun as shown in Figure 2.6. On the following illustration, complete the labeling by identifying wavelengths, naming the portions of the spectrum, and identifying the components of visible light.

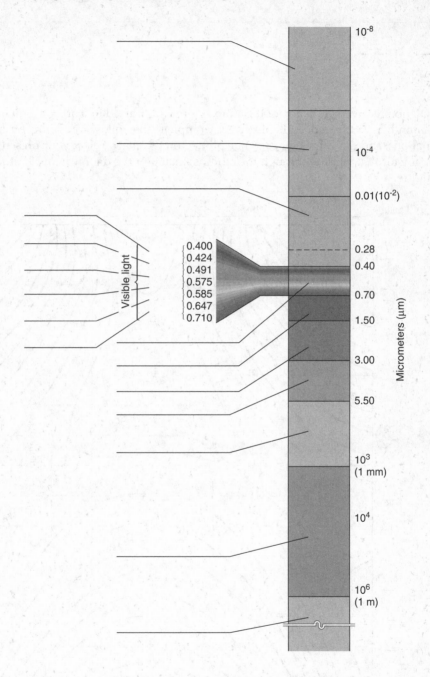

2. What is a wavelength? Give a couple of examples of different wavelengths. _____

_____

_____

_____

3. What wavelength does the Sun emit in the greatest amount? What wavelength does Earth emit in the greatest amount? In which portions of the electromagnetic spectrum do these wavelengths fall? What physical objects on Earth emit energy close to these wavelengths? _____

_____

_____

_____

4. Total daily insolation received at the top of the atmosphere and charted in watts per square meter per day by latitude and month is presented in Figure 2.10. Complete the following figure by adding months, latitudinal designations, and the values for each line plotted on the chart. Select your present latitude on the chart and draw a line for that latitude through the months, then plot the data for your latitude on the graph that follows the figure.

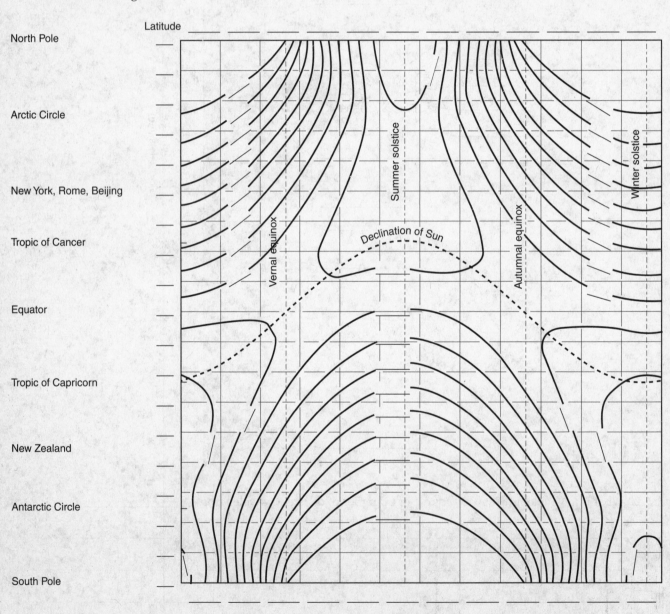

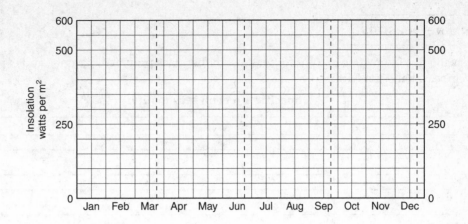

(a) Graph of energy received at your latitude. (See the four graphs as examples in Figure 2.10.)

5. Describe the intercepted energy at the top of the atmosphere. Include discussion of the *thermopause, insolation, solar constant,* and *subsolar point* on Earth's surface below. _____

_____

_____

6. Why is the input of energy into the Earth–atmosphere system unevenly received by latitude? Explain Figure 2.9. _____

_____

7. Figure 2.11 presents average daily net radiation flows at the top of the atmosphere. In Chapter 4, Figure 4.13 summarizes energy surpluses and deficits for the entire Earth–atmosphere system. How do these two figures tie together? What do they tell you about Earth's energy budget? What can you infer about the patterns of climate and vegetation at the surface produced by these uneven energy receipts?

_____

_____

_____

_____

8. On Figure 2.11, Daily Net Radiation Patterns, find your approximate location (where you are now reading these words). What is the average daily net radiation at this location at the top of the atmosphere (units are W/m^2)? _____

✳ STEP 5: The Seasons

1. Table 2.1 lists the five interacting reasons for seasons. In the space provided, <u>describe</u> and <u>explain</u> in what way each affects seasonality on Earth.

Factor: **Specific effect on seasonality:**

Factor	Specific effect on seasonality
a) _____	_____
b) _____	_____
c) _____	_____
d) _____	_____
e) _____	_____

2. <u>Summarize</u> the following key seasonal anniversary dates from Tables 2.3 and 2.4.

Approx. date	Name	Subsolar point location (Sun's declination)	Daylength at 40° North
December			
March			
June			
September			

3. <u>Describe</u> Earth's rotation on its axis as completely as possible. _____
_____
_____

4. For your latitude and location, give approximate sunrise and sunset times (go to the sunrise/sunset calculator at **http://www.srrb.noaa.gov/highlights/sunrise/sunrise.html**; or use Table 2.3 for estimating, or use a local information source):

      December 21–22:   sunrise: _____   sunset: _____

      March 20–21:   sunrise: _____   sunset: _____

      June 20–21:   sunrise: _____   sunset: _____

      September 22–23:   sunrise: _____   sunset: _____

5. Using Figure 2.17 for 50° north latitude, describe the Sun's *altitude* for the following key dates:

      December 21–22:   _____

      March 20–21:   _____

      June 20–21:   _____

      September 22–23:   _____

6. Have you noticed these seasonal changes in Sun altitude through a window, in the backyard, or with the orientation of your house or apartment? Describe.

_____

_____

_____

7. Compare and contrast the two remote sensing images in Figure 2.18. What principal differences in seasonal change do you notice between January and July?

_____

_____

✳ STEP 6: NetWork—Internet Connection

There are many Internet addresses (URLs) listed in this chapter of the *Geosystems* textbook. Go to any two URLs, or "Destination" links on the *Geosystems* Home Page, and briefly describe what you find.

1. _____ : _____

2. _____ : _____ :

_____

SAMPLE SELF-TEST

(Answers appear at the end of the study guide.)

1. Our planet and our lives are powered by

    a. energy derived from Earth's systems
    b. radiant energy from the star closest to Earth
    c. utilities and oil companies
    d. the solar wind

2. In terms of its distance from the Sun, Earth is

    a. the farthest planet of the nine
    b. the closest planet of the nine
    c. about 50,000 km (30,000 mi) away
    d. closer in January and farther in July

3. The Sun produces which of the following?

    a. visible light only
    b. streams of charged particles and radiant energy
    c. only the solar wind
    d. only radiant energy

4. The auroras in the upper atmosphere are caused by

    a. the interaction of electromagnetic radiant energy with atmospheric gases
    b. Earth-generated radio broadcasts
    c. various weather phenomena
    d. the interaction of the solar wind and atmospheric gases

5. Intercepted solar radiation is called

    a. solar wind
    b. light
    c. thermosphere
    d. insolation

6. The Sun is

    a. one of the largest stars in the Milky Way Galaxy
    b. a star of medium size and temperature, about average for stars
    c. a star that produces only energy beneficial to life on Earth
    d. a giant planet

7. Which of the following is correct?

    a. insolation is more diffuse at the equator
    b. insolation is more concentrated at higher latitudes
    c. Earth's sphericity has no effect on insolation received at the surface
    d. insolation is more concentrated at the equator and diffuse at higher latitudes

8. Changes in *daylength* and the Sun's *altitude* represent

    a. revolution
    b. phenomena that occur only at the equator
    c. the concept of seasonality
    d. factors that remain constant and unchanging throughout the year

9. The *plane of the Earth's orbit* about the Sun is called

    a. the plane of the ecliptic
    b. perihelion
    c. aphelion
    d. a great circle

10. The Sun's declination refers to

    a. the angular distance from the equator to the point where direct overhead insolation is received (the subsolar point)
    b. the angular difference between the horizon and the Sun
    c. the Sun's tilt on its axis
    d. its altitude in the sky

11. The Tropic of Capricorn refers to

    a. the parallel that is 23.5° south latitude
    b. the location of the subsolar point on September 22
    c. the parallel at the farthest northern location for the subsolar point during the year
    d. the location of the subsolar point during the vernal equinox

12. On June 21st, the Sun's declination is at

    a. the equator
    b. Rio de Janeiro, Brazil, and Alice Springs, Australia
    c. the Tropic of Capricorn
    d. the Tropic of Cancer

13. The Solar System is located on a remote trailing edge of the Milky Way Galaxy.

    a. true
    b. false

14. It takes sunlight 24 hours to reach the top of the atmosphere from the Sun at the speed of light.

    a. true
    b. false

15. The average amount of energy received at the thermopause is 1372 watts per m^2 (2 cal/cm^2/min).

    a. true
    b. false

16. The Sun emits radiant energy that is composed of 8% _____

_____wavelengths; 47% _____

wavelengths; and 45% _____wavelengths.

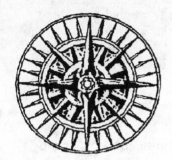

3

EARTH'S MODERN ATMOSPHERE

Earth's atmosphere is a unique reservoir of gases, the product of billions of years of evolutionary development. This chapter examines the atmosphere's structure, function, and composition. Insolation arrives from outer space and descends through the various layers and regions of the atmosphere.

The guiding concept of this chapter involves the flow of insolation from the top of the atmosphere down through the atmosphere to Earth's surface. A consideration of our modern atmosphere must also include the spatial aspects of human-induced gases that affect it, such as air pollution, the stratospheric ozone predicament, and the blight of acid deposition.

The modern atmosphere is an essential medium for life processes. We feel its significance during every Space Shuttle flight and witness the equipment and safeguards needed to protect humans when they venture outside the atmosphere's life-sustaining and protective cover.

An interesting thought exercise is to take the idea of an astronaut doing a spacewalk (EVA). Refer to the photo in the Chapter 1 Career Link of astronaut Dr. Thomas Jones (Figure 2), or the *Apollo* astronaut in Figure 2.3, Chapter 2, standing on the Moon's surface. If you were in charge of designing and constructing the spacesuit, your design would have to include the elements in Figure 3.2, be self-contained with all the functions of the biosphere, and accomplish the protection shown in Figure 3.6.

OUTLINE HEADINGS AND KEY TERMS

The first-, second-, and third-order headings that divide Chapter 3 serve as an outline for your notes and studies. The key terms and concepts that appear **boldface** in the text are listed here under their appropriate heading in ***bold italics***. All these highlighted terms appear in the text glossary. Note the check-off box (❏) so you can mark your progress as you master each concept. These terms should be in your reading notes or used to prepare note cards. The ✪ icon indicates that there is an accompanying animation on the Student CD. The ✿ icon indicates that there is an accompanying satellite or notebook animation on the CD.

The outline headings and terms for Chapter 3:

Atmospheric Composition, Temperature, and Function

✪ **Ozone Breakdown, Ozone Hole**
- ❏ *exosphere*

Atmospheric Profile
- ❏ *air pressure*

Atmospheric Composition Criterion

Heterosphere
- ❏ *heterosphere*

Homosphere
- ❏ *homosphere*

Atmospheric Temperature Criterion

Thermosphere
- ❏ *thermosphere*
- ❏ *thermopause*
- ❏ *kinetic energy*
- ❏ *heat*
- ❏ *sensible heat*

Mesosphere
- ❏ *mesosphere*
- ❏ *noctilucent clouds*

Stratosphere
- ❏ *stratosphere*

Troposphere

- ❏ *troposphere*
- ❏ *tropopause*
- ❏ *normal lapse rate*
- ❏ *environmental lapse rate*

Atmospheric Function Criterion

Ionosphere

- ❏ *ionosphere*

Ozonosphere

- ❏ *ozonosphere, ozone layer*
- ❏ *chlorofluorocarbon compounds (CFCs)*

Variable Atmospheric Components

Natural Sources

Natural Factors that Affect Air Pollution

Winds

Local and Regional Landscapes

Temperature Inversion

- ❏ *temperature inversion*

Anthropogenic Pollution

Carbon Monoxide Pollution

- ❏ *carbon monoxide (CO)*

Photochemical Smog Pollution

- ❏ *photochemical smog*
- ❏ *peroxyacetyl nitrates (PAN)*
- ❏ *nitrogen dioxide (NO_2)*
- ❏ *volatile organic compounds (VOCs)*

Industrial Smog and Sulfur Oxides

- ❏ *industrial smog*
- ❏ *sulfurdioxide*
- ❏ *sulfate aerosols*

Particulates

- ❏ *particulate matter (PM)*
- ❏ *anthropogenic atmosphere*

Benefits of the Clean Air Act

SUMMARY AND REVIEW

News Reports, Focus Studies, and High Latitude Connection

News Report 3.1: Falling Through the Atmosphere—The Highest Sky Dive

Focus Study 3.1: Stratospheric Ozone Losses: A Worldwide Health Hazard

✿ **Southern Hemisphere Ozone 2002–2003 Satellite Loop**

Focus Study 3.2: Acid Deposition: Damaging to Ecosystems

High Latitude Connection 3.1: Arctic Haze

URLs listed in Chapter 3

Noctilucent clouds:
http://lasp.colorado.edu/noctilucent_clouds/

Mauna Loa CO_2 monitoring:
http://cdiac.esd.ornl.gov/ftp/maunaloa-co2/ maunaloa.co2

Air pollution observations and agreements:
http://www.epa.gov/
http://www.ec.gc.ca/
http://www.ijc.org/

Stratospheric ozone predicament:
http://www.cancer.org/
http://toms.gsfc.nasa.gov/teacher/ozone_overhead.html
http://www.epa.gov/sunwise/uvindex.html
http://www.epa.gov/ozone/index.html
http://www.ec.gc.ca/ozone/indexe.htm
http://www.ec.gc.ca/ozone/tocmontr.htm
http://jwocky.gsfc.nasa.gov/
http://www.unep.org/ozone/index-en.shtml

United Nations Ozone Secretariat:
http://www.unep.org/unep/secretar/ozone/home.htm

Montreal Protocol:
http://www.ec.gc.ca/international/multilat/ozone1_e.htm

KEY LEARNING CONCEPTS FOR CHAPTER 3

The following key learning concepts help guide your reading and comprehension efforts. The operative word is in *italics*. Use these carefully to guide your reading of the chapter and note that STEP 1 asks you to work with these concepts. These same learning concepts are used in organizing the summary and review section at the end of the chapter—grouping together definitions, a list of key terms, and review questions.

After reading the chapter and using this study guide, you should be able to:

- *Construct* a general model of the atmosphere based on the criteria composition, temperature, and function, and *diagram* this model in a simple sketch.
- *List* the stable components of the modern atmosphere and their relative percentage contributions by volume, and *describe* each.
- *Describe* conditions within the stratosphere; specifically, *review* the function and status of the ozonosphere (ozone layer).
- *Distinguish* between natural and anthropogenic variable gases and materials in the lower atmosphere.
- *Describe* the sources and effects of carbon monoxide, nitrogen dioxide, and sulfur dioxide, and *construct* a simple equation that illustrates photochemical reactions that produce ozone, peroxyacetyl nitrates, nitric acid, and sulfuric acid.

✳ STEP 1: Critical Thinking Process

Using your interest and learning, and the following questions as guidelines <u>only</u>, briefly discuss your experience with this chapter. In examining your learning you need not go through each of these questions in detail, simply provide an overview of your critical thinking process as it relates to some aspect of this chapter.

- What did you know about the learning concept before you began?
- Which information sources did you use in your learning (text, class, other)?
- Were you able to complete the action stated in the learning concept? What did you learn?
- Were you able to complete the action stated in the learning concept? What did you learn?
- Are there any aspects of the concept about which you want to know more?

Critical Thinking and Chapter 3: _____

_____

_____

_____

_____

✳ STEP 2: Atmospheric Structure and Composition

1. The text and Figure 3.2 classify the atmosphere according to three criteria. <u>Identify</u> the portion of the atmosphere defined by each criteria and record their physical extent in km (or mi), starting at Earth's surface and moving upward to the top of the atmosphere, in the spaces provided.

Criteria	Name for atmospheric region	Extent in km (mi)
Composition:	**a)**	
	b)	
Temperature:	**a)**	
	b)	
	c)	
	d)	
Function:	**a)**	
	b)	

2. List the four principal gases of the modern homosphere by name, % by volume, and parts per million.

a) _____

b) _____

c) _____

d) _____

3. Relative to atmospheric air pressure:

a) What is normal sea-level pressure expressed in mb, in., and kilopascals?

_____

b) Air pressure decreases with altitude (Figure 3.3a and b). According to the text, how much of the atmosphere is compressed <u>below</u> each of the following altitudes? Give the percent of atmosphere that exists below that altitude.

5500 m (18,000 ft): _____

10,700 m (35,100 ft): _____

16,000 m (52,500 ft): _____

50 km (31 mi): _____

4. Relative to News Report 3.1, "Falling Through the Atmosphere—The Highest Sky Dive," what did Captain Kittinger find initially alarming? Which layers of the atmosphere did he fall through in terms of composition, temperature, and function? _____

_____

_____

5. Figure 3.2 is an integrated illustration that identifies the criteria for classification of the atmosphere's temperature profile with altitude, as well as other information. Use the reproduction of Figure 3.2 below to record all the elements and complete the labels as the figure appears in the text. You can add information, labels, and coloration (colored pencils are best) as you read through the text.

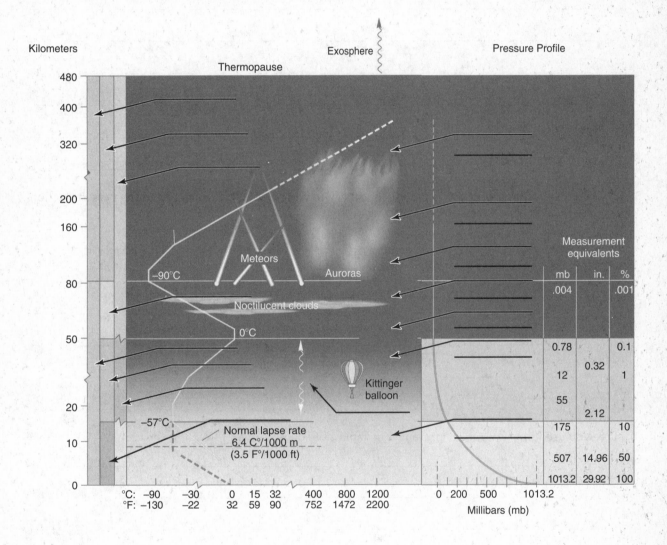

✳ STEP 3: Stratospheric Ozone Losses

1. Describe the current problems relative to stratospheric ozone, the identification of the problem and its causes, and a brief history of the science, current action being taken, and the present status of the ozonosphere.

_____

_____

_____

_____

_____

_____

2. What is your interpretation of Figure 3.1.2a and b in Focus Study 3.1? Explain.

_____

_____

_____

_____

3. What does the depletion of stratospheric ozone mean to you personally? Do you have any personal action or response plans related to this situation? Any products you buy?

_____

_____

_____

_____

4. Have you used the "UV Index" presented in Table 1 of Focus Study 3.1? What is the current "UV Index"

where you live? _____

_____

_____

_____

✳ STEP 4: Variable Atmospheric Components

1. From Figures 3.7 and 3.8, and Table 3.3, describe some natural variable sources of air pollution. _____

_____

_____

2. What are the sources of Arctic haze and where are they located? What does this tell us about how pollution spreads through the atmosphere?

_____

_____

_____

3. Using Figures 3.12 and 3.13, Table 3.4, and the text, describe how each of the following photochemical pollutants is formed in the lower troposphere. Simple generalizations of the formula in the illustration are all right. Assume that the source of nitrogen dioxide is from transportation systems.

a) Ozone (O_3): _____

_____

b) PAN (peroxyacetyl nitrates): _____

_____

c) Nitric acid (HNO_3): _____

_____

4. What is acid deposition? Why a "blight on the landscape" (Focus Study 3.2)? What processes have scientists identified that lead to the formation of acids in the atmosphere? Include in your writing a description of what is portrayed in the pair of photos in Figure 3.2.1 of the focus study.

_____

_____

_____

What solutions do you recommend? _____

_____

_____

5. The last heading in the chapter explains "Benefits of the Clean Air Act." What were the actual benefits accrued to the public from 1970 to 1990 from clean-air regulations? Briefly summarize the financial impact of the Clean Air Act. Address this question: Can you make an argument for repeal or weakening of the Clean Air Act?

_____

_____

_____

✳ STEP 5: NetWork—Internet Connection

There are many Internet addresses (URLs) listed in this chapter of the _Geosystems_ textbook. Go to any two URLs, or "Destination" links on the _Geosystems_ Home Page, and briefly describe what you find.

1. _____ : _____

_____

2. _____ : _____

_____

SAMPLE SELF-TEST
(Answers appear at the end of the study guide.)

1. Lewis Thomas, in his book, _The Lives of the Cell_, compared Earth's atmosphere with a

    a. layer of human skin
    b. membrane of a cell
    c. layer of water
    d. magnetosphere

2. Which of the following is true?

    a. three-fourths of the atmosphere occurs below 10,700 m (35,105 ft)
    b. 90% of the atmosphere is below 5,500 m (18,000 ft)
    c. 90% of the atmosphere remains above the tropopause
    d. half of the total atmosphere occurs below the stratopause

3. Based on <u>composition</u>, the atmosphere is divided into

    a. five regions beginning with the outermost thermosphere
    b. two broad regions
    c. two functional areas that absorb radiation from the Sun
    d. the troposphere and the stratosphere

4. The region of the atmosphere that is so evenly mixed that it behaves as if it were a single gas is

    a. the homosphere
    b. the heterosphere
    c. the exosphere
    d. the thermosphere

5. Which of the following lists of gases is correct, from <u>most to least</u> in terms of percentage within the homosphere?

   a. nitrogen, argon, oxygen, xenon, carbon dioxide
   b. nitrogen, oxygen, argon, carbon dioxide, trace gases
   c. oxygen, PAN, ozone, nitrogen, carbon dioxide
   d. water vapor, oxygen, argon, carbon dioxide

6. Recent measurements of increased levels of ultraviolet light at Earth's surface

   a. are focused along the equator and not the polar regions
   b. are related to an increasing rate of skin cancer that is rising at 10% per year
   c. are unrelated to stratospheric ozone
   d. affect those at sea level more than those living in mountains

7. Directly above the midlatitudes, the tropopause ($-57°C$, $-70°F$) occurs at approximately

   a. the stratopause
   b. 1 km
   c. 13 km (8 mi)
   d. 80 km (50 mi)

8. Relative to lapse rates in the troposphere,

   a. the <u>environmental</u> lapse rate refers to the actual lapse rate at any particular time and may vary greatly from the normal lapse rate
   b. temperatures tend to increase with altitude
   c. temperatures remain constant with increasing altitude
   d. an average normal lapse rate value of 10 C° per 1000 m increases with altitude

9. Which of the following is false relative to carbon dioxide?

   a. it is critically important in regulating the temperature of the planet
   b. it occurs at nearly 0.037% in the lower atmosphere
   c. the amount of carbon dioxide has <u>increased</u> as a result of human activities
   d. all of the above are true statements; none of these is false

10. PAN in the lower troposphere

   a. is principally related to sulfur dioxides
   b. is formed by particulates such as dust, dirt, soot, and ash
   c. damages and kills plant tissue, a photochemical product
   d. comes directly from automobile exhaust

11. Industrial smog is

   a. associated with photochemistry
   b. principally associated with coal-burning industries
   c. a relatively recent problem during the latter half of this century
   d. principally associated with transportation

12. The oxides of sulfur and nitrogen

   a. lead to the formation of airborne sulfuric and nitric acid
   b. form acids that are deposited in both dry and wet forms
   c. are produced by industry and transportation
   d. all of these choices are true

13. Variable atmospheric components refer to

   a. only natural gases and materials
   b. nitrogen, oxygen, argon, and carbon dioxide
   c. volcanic dust, forest fire smoke, but nothing anthropogenic (human-caused)
   d. both natural and anthropogenic gases and materials

14. A temperature inversion occurs

    a. when surface temperatures are higher than overlying layers of air
    b. when there is good air drainage and ventilation of the surface air
    c. when surface temperatures are lower than warmer overlying air
    d. during episodes of reduced air pollution

15. The two layers of the atmosphere specifically defined according to their <u>function</u> are the troposphere and the heterosphere.

    a. true
    b. false

16. In the lower atmosphere, nitrogen is essentially a by-product of photosynthesis, whereas oxygen is a product of bacterial action.

    a. true
    b. false

17. The ozonosphere is presently under attack by photochemical smog.

    a. true
    b. false

18. Weather occurs primarily in the troposphere.

    a. true
    b. false

19. The principal gases in the atmosphere are nitrogen, carbon dioxide, and argon.

    a. true
    b. false

20. Arctic haze is caused by volcanic activity in the Arctic.

    a. true
    b. false

21. Both nitrogen dioxide and sulfur dioxide are principally produced by automobiles.

    a. true
    b. false

4

ATMOSPHERE AND SURFACE ENERGY BALANCES

Earth's biosphere pulses with flows of energy. Chapter 4 follows the passage of solar energy through the lower atmosphere to Earth's surface. This entire process is a vast flow-system with energy cascading through fluid Earth systems. An integrated illustration, Figure 4.12, presents the Earth–atmosphere energy balance; referencing this figure as you read the chapter is helpful—a simplified version is in Figure 4.1. The chapter then analyzes surface energy budgets and develops the concept of net radiation and the expenditure pathways for this energy. In addition, the energy characteristics of urban areas are explored, for the climates in our cities differ measurably from those of surrounding rural areas.

New remote-sensing capabilities are helping scientists understand more about Earth's energy budgets. Figure 4.8 looks at aerosols, including black carbon, that both reflect and absorb incoming insolation as identified in four images by the CERES sensors aboard satellite *Terra* between January and March 2001, for southern Asia and the Indian Ocean. The increased aerosol levels from human activities produce higher albedos and atmospheric absorption of energy, resulting in a lowering of surface temperatures.

Figure 4.11 from the CERES sensors aboard *Terra* made these portraits in March 2000, which show outgoing shortwave energy flux reflected from clouds, land, and water—Earth's albedo— and longwave energy flux emitted by surfaces back to space.

OUTLINE HEADINGS AND KEY TERMS

The first-, second-, and third-order headings that divide Chapter 4 serve as an outline for your notes and studies. The key terms and concepts that

appear **boldface** in the text are listed here under their appropriate heading in ***bold italics***. All these highlighted terms appear in the text glossary. Note the check-off box (❏) so you can mark your progress as you master each concept. These terms should be in your reading notes or used to prepare note cards. The ✪ icon indicates that there is an accompanying animation on the Student CD. The ✹ icon indicates that there is an accompanying satellite or notebook animation on the Cd.

The outline headings and terms for Chapter 4:

Energy Essentials
- ✪ ***Global Warming, Climate Change***
- ✹ ***Global Albedo Values Satellite Loop***
- ✹ ***Global Shortwave Radiation Satellite Loop***

Energy Pathways and Principles

  ❏ ***transmission***

Insolation Input

Scattering (Diffuse Radiation)

  ❏ ***scattering***
  ❏ ***diffuse radiation***

Refraction

  ❏ ***refraction***

Albedo and Reflection

  ❏ ***reflection***
  ❏ ***albedo***

Clouds, Aerosols, and the Atmosphere's Albedo

  ❏ ***cloud-albedo forcing***
  ❏ ***cloud-greenhouse***
  ❏ ***forcing***

Absorption

❑ *absorption*

Conduction, Convection, and Advection

❑ *conduction*
❑ *convection*
❑ *advection*

Energy Balance in the Troposphere

✪ **Earth–Atmosphere Energy Balance**

The Greenhouse Effect and Atmospheric Warming

❑ *greenhouse effect*

Clouds and Earth's "Greenhouse"

Earth–Atmosphere Radiation Balance

Energy Balance at Earth's Surface

✸ **Global Net Radiation Satellite Loop**
✸ **Global Latent Heat Flux Satellite Loop**
✸ **Global Sensible Heat Satellite Loop**

Daily Radiation Patterns
Simplified Surface Energy Balance

❑ *microclimatology*
❑ *net radiation (NET R)*

Net Radiation

Sample Stations

The Urban Environment

❑ *urban heat island*
❑ *dust dome*

SUMMARY AND REVIEW

News Report and Focus Study

News Report 4.1: Earthshine Studies—Possible Energy Budget Diagnostic
Focus Study 4.1: Solar Energy Collection and Concentration

URLs listed in Chapter 4

Asian energy study and CERES:
http://www-indoex.ucsd.edu/
http://asd-www.larc.nasa.gov/ceres/ASDceres.html

NASA urban heat island studies and heat island images:
http://www.ghcc.msfc.nasa.gov/urban/
http://science.nasa.gov/newhome/headlines/ essd01jul98%5F1.htm

Solar Cookers International:
http://www.solarcooking.org/

National Renewable Energy Laboratory (NREL):
http://rredc.nrel.gov/solar/pubs/NSRDB/
http://www.nrel.gov
http://www.nrel.gov/ncpv/
http://www.eren.doe.gov/pv/

KEY LEARNING CONCEPTS FOR CHAPTER 4

The following key learning concepts help guide your reading and comprehension efforts. The operative word is in *italics*. Use these carefully to guide your reading of the chapter and note that STEP 1 asks you to work with these concepts. These same learning concepts are used in organizing the summary and review section at the end of the chapter—grouping together definitions, a list of key terms, and review questions.

After reading the chapter and using this study guide, you should be able to:

- *Identify* the pathways of solar energy through the troposphere to Earth's surface: transmission, refraction, albedo (reflectivity), scattering, diffuse radiation, conduction, convection, and advection.
- *Describe* what happens to insolation when clouds are in the atmosphere, and *analyze* the effect of clouds and air pollution on solar radiation received at ground level.
- *Review* the energy pathways in the Earth–atmosphere system, the greenhouse effect, and the patterns of global net radiation.
- *Plot* the daily radiation curves for Earth's surface, and *label* the key aspects of incoming radiation, air temperature, and the daily temperature lag.
- *Portray* typical urban heat island conditions, and *contrast* the microclimatology of urban areas with that of surrounding rural environments.

✳ STEP 1: Critical Thinking Process

Using your interest and learning, and the following questions as guidelines only, briefly discuss your experience with this chapter. In examining your learning you need not go through each of these questions in detail, simply provide an overview of your critical thinking process as it relates to some aspect of this chapter.

- What did you know about the learning concept before you began?
- Which information sources did you use in your learning (text, class, other)?
- Were you able to complete the action stated in the learning concept? What did you learn? Are there any aspects of the concept about which you want to know more?

Critical Thinking and Chapter 4: _____

_____

_____

_____

✳ STEP 2: Energy in the Troposphere

1. Relative to albedo: in the visible wavelengths, _____ colors have lower albedos, and _____ colors have higher albedos. On water surfaces, the angle of the solar rays also affects albedo values; _____ angles produce a greater reflection than do _____ angles. In addition, _____ surfaces increase albedo, whereas _____ surfaces reduce it.

2. Record albedo values (percentages reflected) for various surfaces in the spaces below, using Figure 4.5.

Fresh snow:	_____
Snow, several days old:	_____
Forests:	_____
Crops, grasslands:	_____
Grass:	_____
Concrete:	_____
Asphalt, blacktop:	_____
Dark house roof:	_____
Light house roof:	_____
Moon surface:	_____
Earth (average):	_____

3. What is earthshine? What can it tell us about Earth's albedo?

_____

_____

_____

4. What is refraction? How does it affect daylength? Explain (Figure 4.4).

_____

_____

5. In terms of cloud cover and energy patterns describe the following (Figure 4.8):

(a) cloud-albedo forcing: _____

_____

(b) cloud-greenhouse forcing: _____

_____

6. What effect do jet contrails have on atmospheric and surface temperatures? How was this studied?

_____

_____

7. Figure 4.12 illustrates the Earth–atmospheric energy balance. The outline of this integrated figure appears below. As you read through the text, <u>fill in</u> the pathways and circuits of energy transmission in the lower atmosphere, the appropriate labels, and the units of energy involved. Please use colored pencils to shade in appropriate areas that best suggest the nature of each pathway.

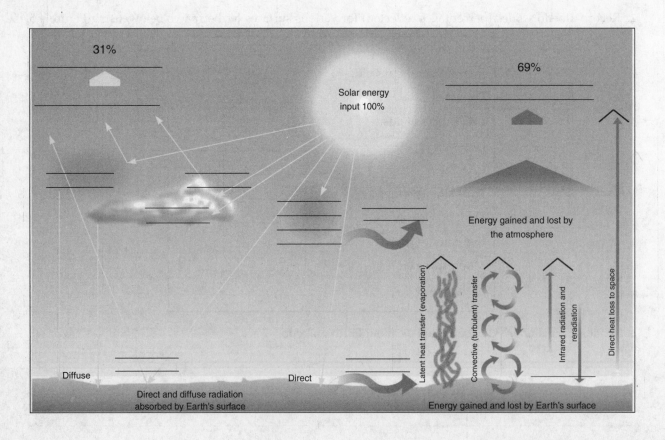

8. In the space provided below, <u>sketch</u> your own version of Earth's global net radiation after Figure 4.13. Properly label the latitudes that exhibit surpluses and deficits of radiation, and the poleward transport of energy surpluses.

✳ STEP 3: Energy at Earth's Surface

1. In the space provided <u>record</u> a simplified surface energy budget, and <u>identify</u> each of the components.

_____

_____

2. Net radiation (NET R) is expended in <u>three</u> general output paths. <u>List</u> these expenditures with their letter symbol and briefly <u>explain</u> each.

(a) _____   _____

_____

(b) _____   _____

_____

(c) _____   _____

_____

3. Figure 4.18 shows the global patterns of the latent heat of evaporation, and Figure 4.19 shows the global pattern of sensible heat. How are the areas of high latent heat and high sensible heat related to each other? How do they relate to the pattern of global net radiation patterns, shown in Figure 4.17? Are there areas with high latent heat and high sensible heat? If so, where is the "extra" energy coming from?

_____

_____

_____

_____

4. Figures 4.20a and c are two examples of daily surface energy balances. What do these balances demonstrate about the climates of the two locations? Which of the components vary the most between the two surface energy graphs?

(a) El Mirage, CA, as a desert, subtropical location: _____

_____

(b) Pitt Meadows, B.C., as a vegetated, moist, midlatitude location: _____

_____

5. With this analysis in mind for El Mirage and Pitt Meadows, compare and analyze the Sahara region of North Africa in terms of latent heat (world map in Figure 4.18) and sensible heat (world map in Figure 4.19). What do you see? Describe.

_____

_____

_____

6. Focus Study 4.1 examines solar energy collection and concentration as a viable energy source. The photographs in Figures 4.1.1b and 4.1.1c demonstrate simple solar cookers. After reading the focus study, the caption, and examining the photos, describe in spatial terms the impact of these low-technology devices on the livelihood patterns of the people pictured.

_____

_____

_____

7. Explain the five characteristics that contribute to urban microclimates. Utilize information from Tables 4.1 and 4.2, and Figures 4.21 and 4.22 in your explanations.

(a) _____

_____

(b) _____

_____

(c) _____

_____

(d) _____

_____

(e) _____

_____

✳ STEP 4: NetWork—Internet Connection

There are many Internet addresses (URLs) listed in this chapter of the *Geosystems* textbook. Go to any two URLs, or "Destination" links on the *Geosystems* Home Page, and briefly describe what you find.

1. _____ : _____

_____

2. _____ : _____

_____

SAMPLE SELF-TEST

(Answers appear at the end of the study guide.)

1. Shortwave and longwave energy passing through the atmosphere or water does so by

   a. absorption
   b. transmission
   c. refraction
   d. insolation
   e. reflection

2. The insolation received at Earth's surface is

   a. usually greatest at the equator
   b. generally greater at high latitudes
   c. greatest over low-latitude deserts with their cloudless skies
   d. usually lowest along the equator because of cloud cover
   e. inadequate to sustain life

3. Heat energy (infrared) is mainly absorbed in the "greenhouse effect" by which two gases?

   a. oxygen and hydrogen
   b. ozone and dust
   c. nitrogen and oxygen
   d. water vapor and carbon dioxide

4. The reflective quality of a surface is known as its

   a. conduction
   b. absorption
   c. albedo

    d. scattering

    e. wavelength

5. Which of the following has the lowest albedo?

    a. Earth, surface average

    b. snow, polluted, several days old

    c. dry, light, sandy soils

    d. forests

    e. the Moon's surface in full sunlight

6. The assimilation of radiation by a surface and its conversion from one form to another is termed

    a. reflection

    b. diffuse radiation

    c. absorption

    d. transmission

    e. refraction

7. Daily temperatures usually

    a. reach a high at the time of the noon Sun

    b. experience a low in the middle of the night

    c. record a high that lags several hours behind the noon Sun

    d. show no relationship to insolation input

8. The blue color of Earth's lower atmosphere is produced by a phenomenon known as scattering.

    a. true

    b. false

9. The quantity of energy mechanically transferred from the air to the surface and surface to air is

    a. latent heat energy

    b. sensible heat transfer

    c. ground heating

    d. net radiation

10. Urban heat islands experience climatic effects related to their

    a. artificial surfaces

    b. lower albedo values

    c. irregular geometric shapes and angles

    d. human occupation and energy conversion systems

    e. all of these are correct

11. Which of the following climatic factors decreases as a result of urbanization

    a. clouds and fog

    b. annual mean temperatures

    c. the presence of condensation nuclei

    d. days with precipitation

    e. relative humidity measurements

12. The highest albedo value listed in Figure 4.5 is _____ for _____ ;

    the lowest albedo value listed is _____ for _____ .

13. List the radiatively active gases that produce Earth's greenhouse. _____

    _____

14. The study of climate at or near Earth's surface is called _____

15. On the average clouds reflect 21% of incoming insolation.

    a. true
    b. false

16. Contrails from jet aircraft reflect energy before it enters the atmosphere, producing a net cooling effect.

    a. true
    b. false

17. Air moves physically from one place to another by the process of conduction.

    a. true
    b. false

18. Net radiation at Earth's surface is a product of all incoming and outgoing shortwave and longwave radiation.

    a. true
    b. false

19. The highest net radiation occurs north of the equator in the Arabian Sea at 185 watts per m^2.

    a. true
    b. false

20. Commercially produced electricity from solar energy conversion has not yet been achieved according to the text.

    a. true
    b. false

21. In urban areas, nighttime minimum temperatures are usually lower on calm, cloudy nights.

    a. true
    b. false

22. Pitt Meadows, B.C., experiences higher latent heat exchanges at the surface than does the desert location of El Mirage, CA.

    a. true
    b. false

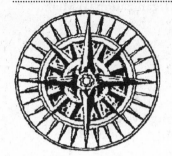

5

GLOBAL TEMPERATURES

Air temperature has a remarkable influence upon our lives, both at the microlevel and at the macrolevel. A variety of temperature regimes worldwide affect entire lifestyles, cultures, decision making, and resources spent. Global temperature patterns appear to be changing in a warming trend that is affecting us all and is the subject of much scientific, geographic, and political interest. Our bodies sense temperature and subjectively judge comfort, reacting to changing temperatures with predictable responses. These are some of the topics addressed in Chapter 5 and in later chapters.

The new chapter-opening photo of a Canadian Prairie farm in Alberta, in autumn in sunset light, brings to mind the record-breaking temperatures (1998 through 2001) that are hitting Canada and all northern latitudes. The increased temperatures are contributing to a drought throughout the region.

This chapter presents concepts that are synthesized on three temperature maps: January, July, and the annual range. As the chapter develops, reference to these maps will be useful. The chapter relates these temperature concepts directly to you with a discussion of apparent temperatures—the wind chill and heat index charts. Focus Study 5.1 introduces some essential temperature concepts to begin our study of world temperatures.

OUTLINE HEADINGS AND KEY TERMS

The first-, second-, and third-order headings that divide Chapter 5 serve as an outline for your notes and studies. The key terms and concepts that appear **boldface** in the text are listed here under their appropriate heading in ***bold italics***. All these highlighted terms appear in the text glossary. Note the check-off box (❑) so you can mark your progress as you master each concept. These terms should be in your reading notes or used to prepare note cards.

The ⚙ icon indicates that there is an accompanying animation on the Student CD. The ✸ icon indicates that there is an accompanying satellite or notebook animation on the Cd.

The outline headings and key terms for Chapter 5:

Temperature Concepts and Measurement

   ❑ ***temperature***

Temperature Scales

Measuring Temperature

Principal Temperature Controls

Latitude

Altitude

Cloud Cover

Land–Water Heating Differences

   ❑ ***land–water heating differences***

Evaporation

Transparency

   ❑ ***transparency***

Specific Heat

   ❑ ***specific heat***

Movement

Ocean Currents and Sea-Surface Temperatures

   ❑ ***Gulf Stream***

Summary of Marine Effects vs. Continental Effects

   ❑ ***marine effect***
   ❑ ***continental effect***

Earth's Temperature Patterns

❋ **Global Sea-surface Temperatures Satellite Loop**

❋ **Global Surface Temperatures, Land and Ocean Satellite Loop**

   ❏ *isotherm*

January Temperature Map

   ❏ *thermal equator*

July Temperature Map

Annual Temperature Range Map

Global Temperatures Suggest a Greenhouse Warming

SUMMARY AND REVIEW

Focus Study, Career Link, and High Latitude Connection

☉ **Global Warming, Climate Change**

Focus Study 5.1: Air Temperature and the Human Body
Career Link 5.1: Dr. Louwrens Hacquebord, Professor of Arctic and Antarctic Studies

High Latitude Connection 5.1: Overview of Trends in the Polar Regions

---

URLs listed in Chapter 5

World Meteorological Organization (WMO):
http://www.wmo.ch/

Cloud climatology:
http://isccp.giss.nasa.gov/
http://www.wmo.ch/web/wcrp/wcrp-home.html

Global Climate Observing System:
http://www.wmo.ch/web/gcos/gcoshome.html

Tropical Ocean Global Atmosphere (TOGA) and Coupled Ocean-Atmosphere Response Experiment (COARE):
http://lwf.ncdc.noaa.gov/oa/coare/index.html

Sea-surface temperatures:
http://podaac.jpl.nasa.gov/
http://eosdismain.gsfc.nasa.gov/eosinfo/Welcome/

Applied temperature concepts:
http://www.hpc.ncep.noaa.gov/heat_index.shtml
http://www.msc-smc.ec.gc.ca/
http://205.156.54.206/om/windchill/index.shtml#calculator

KEY LEARNING CONCEPTS FOR CHAPTER 5

The following key learning concepts help guide your reading and comprehension efforts. The operative word is in *italics*. Use these carefully to guide your reading of the chapter and note that STEP 1 asks you to work with these concepts. These same learning concepts are used in organizing the summary and review section at the end of the chapter—grouping together definitions, a list of key terms, and review questions.

   After reading the chapter and using this study guide, you should be able to:

- *Define* the concepts of temperature, kinetic energy, and sensible heat, and *distinguish* among Kelvin, Celsius, and Fahrenheit scales and how they are measured.
- *List* and *review* the principal controls and influences that produce global temperature patterns.
- *Review* the factors that produce marine effects and continental effects as they influence temperatures, and *utilize* several pairs of stations to illustrate these differences.
- *Interpret* the pattern of Earth's temperatures from their portrayal on January and July temperature maps and on a map of annual temperature ranges.
- *Contrast* wind chill and heat index, and *determine* human response to these apparent temperature effects.

❋ **STEP 1: Critical Thinking Process**

Using your interest and learning, and the following questions as guidelines <u>only</u>, briefly discuss your experience with this chapter. In examining your learning you need not go through each of these questions in detail, simply provide an overview of your critical thinking process as it relates to some aspect of this chapter.

- What did you know about the learning concept before you began?
- Which information sources did you use in your learning (text, class, other)?
- Were you able to complete the action stated in the learning concept? What did you learn?
- Are there any aspects of the concept about which you want to know more?

Critical Thinking and Chapter 5: _____

_____

_____

_____

_____

✳ STEP 2: Temperature Basics

1. Using the conversion tables in Appendix C, complete the following conversions. Note the position of the degree symbol and the related explanation in Figure 5.1, and that a different conversion method is used depending on degree symbol placement.

25°C _____	31°F _____	29C° _____	−41F° _____
4°C _____	89°F _____	61C° _____	53F° _____
39°C _____	46°F _____	26C° _____	5F° _____
−40°C _____	14°F _____	−5C° _____	12F° _____

2. Relative to atmospheric temperature, see if you can locate a properly installed thermometer either at the college or university you are attending, or perhaps at home. If you do not have access to a thermometer find another source of information for local temperature: a radio or television station, a local cable channel, the Weather Channel, a local newspaper, the National Weather Service or Environment Canada office, or consult most of these sources on the Internet. For at least 3 days, record the air temperature each day at approximately the same time. See if you can detect a trend or relationship of air temperature to weather and other atmospheric phenomena.

Air temperature observations:

Day 1: _____ Place:_____ Time: _____

Day 2: _____ · Place:_____ Time: _____

Day 3: _____ Place:_____ Time: _____

3. Graphing temperature and precipitation information is an important skill in weather and climate analysis. Note the number of such graphs in *Geosystems*. Let's work with two stations: Portland, Oregon, and Chicago, Illinois.

To begin: Locate Portland, Oregon, and Chicago, Illinois, on the January and July average temperature maps (Figures 5.14 and 5.16) and on the annual range of temperature map (Figure 5.17). Determine the following temperatures for each of these cities (in °C and °F) through interpolation using the isotherms on these general, small-scale maps. (This involves a careful "guesstimate" of the temperature values between isotherms.)

Portland, Oregon: Jan _____ July _____ Annual Range _____

Chicago, Illinois: Jan _____ July _____ Annual Range _____

4. Next, using the graphs on the next two pages, plot the temperature (line graph) and precipitation (bar graph) data for both Portland, Oregon, and Chicago, Illinois. Data are provided next to the graphs. Label each temperature plot and precipitation plot and fill in the city information. Finally, use Figure 10.5 to find the climate classification for each station. You may wish to use an atlas to find the locations of the cities.

A *climograph* is a graph on which temperature, precipitation, and other weather information is plotted. The following is an example of how your final climographs should appear.

Climograph example:

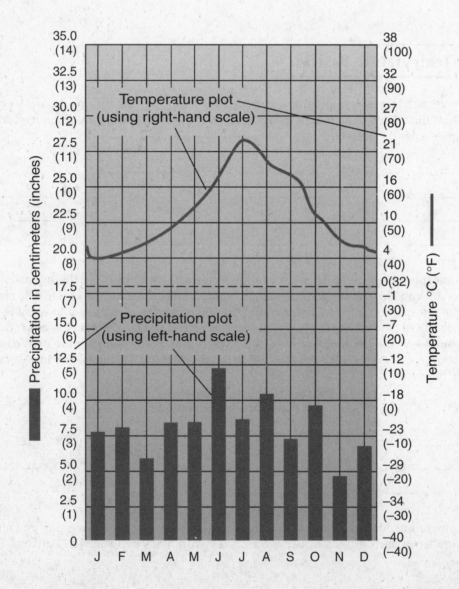

Portland, Oregon
Mediterranean dry, cool summer

Latitude _____

Longitude _____

Elevation _____

Population _____

Total annual rainfall: _____

Average annual temperature: _____

Annual temperature range: _____

Distribution of temperature during the year:

_____

_____

Distribution of precipitation during the year:

_____

_____

_____

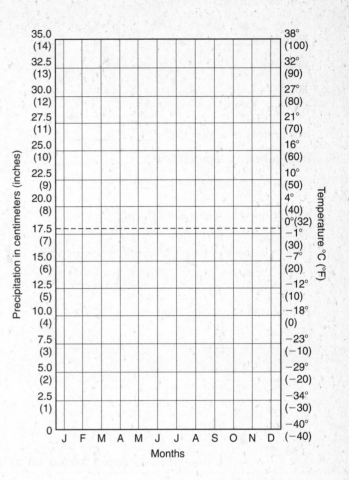

Portland, Oregon: pop. 366,000, lat. 45°31′N, long. 122°40′W, elev. 9 m (30 ft).

	Jan	Feb	Mar	Apr	May	Jun	Jul	Aug	Sep	Oct	Nov	Dec	Annual
Temperature °C	4.4	6.7	8.9	12.2	15.0	17.8	20.0	20.0	17.8	13.3	8.3	5.6	12.8
(°F)	(40.0)	(44.0)	(48.0)	(54.0)	(59.0)	(64.0)	(68.0)	(68.0)	(64.0)	(56.0)	(47.0)	(42.0)	(55.0)
PRECIP cm	13.7	12.4	10.7	6.1	4.8	4.1	1.0	1.5	4.6	8.9	15.2	18.0	101.3
(in.)	(5.4)	(4.9)	(4.2)	(2.4)	(1.9)	(1.6)	(0.4)	(0.6)	(1.8)	(3.5)	(6.0)	(7.1)	(39.9)

Climate classification: _____

_____

Chicago, Illinois

Humid continental, hot summer

Latitude _____

Longitude _____

Elevation _____

Population _____

Total annual rainfall: _____

Average annual temperature: _____

Annual temperature range: _____

Distribution of temperature during the year:

_____

_____

Distribution of precipitation during the year:

_____

_____

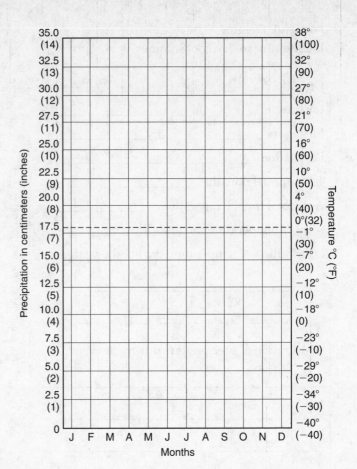

Chicago, Illinois: pop. 3,000,000, lat. 41°47′N, long. 87°35′W, elev. 45 m (610 ft).

	Jan	Feb	Mar	Apr	May	Jun	Jul	Aug	Sep	Oct	Nov	Dec	Annual
Temperature °C	−3.9	−2.8	2.3	8.6	14.4	20.0	23.2	22.4	18.7	12.4	4.6	−1.4	9.9
(°F)	(25.0)	(27.0)	(36.1)	(47.5)	(57.9)	(68.0)	(73.8)	(72.3)	(65.7)	(54.3)	(40.3)	(29.5)	(49.8)
PRECIP cm	4.9	4.8	6.8	7.4	9.0	9.3	8.4	8.0	7.6	6.7	5.9	5.0	83.8
(in.)	(1.9)	(1.9)	(2.7)	(2.9)	(3.5)	(3.7)	(3.3)	(3.1)	(3.0)	(2.6)	(2.3)	(2.0)	(33.0)

Climate classification: _____

_____

5. Can you determine any indication of *marine* effect or *continental* effect influence on the two graphs you just completed? Be specific.

_____

_____

6. Relate your answer in #5 to the temperature graphs in Figures 5.12, 5.13, and 5.15. Compare temperature patterns at the marine cities with those of the continental cities.

_____

_____

_____

_____

7. On a visit to Mount Shasta City (elevation 900 m; 2950 ft) you find the outside air temperature to be 26°C (79°F), and you also find the weather conditions about average in terms of *normal lapse rate* in the lower atmosphere.

Given these conditions, what would the air temperature be (°C and °F) at the summit of Mount Shasta (elevation 4315 m; 14,100 ft) at the same time it is 26°C in the city? (Remember the normal lapse rate is 6.4C°/1000 m times the altitude difference between the city and the summit; use this to calculate the temperature difference.)

Summit temperature: _____ °C _____ °F. Show your work: _____

_____

✳ STEP 3: Global Temperature Maps

1. Using specific references to each continent, what effects do you see on the January (Figure 5.14) and July (Figure 5.16) maps that are caused by higher elevations?

_____

_____

2. Analyze the annual temperature pattern experienced in north central Siberia. Use specifics from the text, the temperature graph for Verkhoyansk (Figure 5.15b), and the three temperature maps for your response.

_____

_____

_____

3. What is meant by the *thermal equator*? Briefly describe its location in the Western Hemisphere over North and South America in January and July. _____

_____

_____

4. High Latitude Connection 5.1 discusses the decrease in Arctic sea ice. What is the decrease in sea ice in the Arctic from 1970 until now? How does the loss in ice affect energy absorption? What kind of feedback is this?

_____

_____

_____

_____

✳ STEP 4: Air Temperature and the Human Response

1. The NWS and the Meteorological Services of Canada (MSC, **http://www.msc-smc.ec.gc.ca/**) revised the wind chill formula and standard assumptions for the 2001–2002 winter season. The new Wind Chill Temperature (WCT) Index is an effort to improve the accuracy of heat loss calculations. Computer modeling, clinical trials and testing, and advances in technology make this revision possible.

Figure 5.1.1 in Focus Study 5.1 is the new version that went into service November 2001, presented in metric with English unit equivalents. The *wind chill* chart presents the *apparent temperature* that you would experience under different temperature and wind conditions (in °C and °F). Determine the wind-chill temperature for each of the following examples (use metric).

(a) Wind speed: <u>24 kmph</u>, air temperature: <u>−34°C</u> = wind chill temp: _____

(b) Wind speed: <u>48 kmph</u>, air temperature: <u>−7°C</u> = wind chill temp: _____

(c) Wind speed: <u>8 kmph</u>, air temperature: <u>+4°C</u> = wind chill temp: _____

2. The *heat index* (HI) (Figure 5.1.2 in Focus Study 5.1) is reported for regions that experience high relative humidity and high temperature readings. This apparent temperature is an important consideration in human health and even survival during extreme HI episodes.

If the relative humidity is 80% and the air temperature is 32.2°C (90°F), then the National Weather Service heat index (HI) rating is a:

Category _____ HI with an apparent temperature of _____ (in °C and °F).

3. According to Focus Study 5.1 and the photo in Figure 5.1.3, what happened in Chicago in 1995?

_____

_____

4. News Report 5.1 discusses the record temperatures that we have experienced over the past two decades. What are some of the possible consequences of this warming? (You might want to flip ahead to Chapter 10 and preview the section on "Global Climate Change.")

_____

_____

✳ STEP 5: NetWork—Internet Connection

There are many Internet addresses (URLs) listed in this chapter of the *Geosystems* textbook. Go to any two URLs, or "Destination" links on the *Geosystems* Home Page, and briefly describe what you find.

1. _____ : _____

2. _____ : _____

SAMPLE SELF-TEST

(Answers appear at the end of the study guide.)

1. Air temperature is a measure of the presence of which of the following in the air?

    a. sensible temperature
    b. apparent temperature
    c. relative humidity
    d. sensible heat

2. Relative to latitude and surface energy receipts, which of the following is true?

    a. insolation intensity <u>increases</u> with distance from the subsolar point
    b. daylength is constant across all latitudes
    c. insolation intensity <u>decreases</u> with distance from the subsolar point
    d. seasonal effects <u>increase</u> toward the equator

3. Relative to temperatures, clouds generally

    a. increase temperature minimums and maximums
    b. cover about 10% of Earth's surface at any one time
    c. are moderating influences, acting like insulation
    d. <u>decrease</u> nighttime temperatures and <u>increase</u> daytime temperatures

4. In general more <u>moderate</u> temperature patterns

    a. are created by continentality
    b. are exemplified by Winnipeg and Wichita
    c. indicate maritime influences
    d. occur at altitude

5. The <u>January</u> mean temperature map (Figure 5.11) shows that

    a. isotherms in North America trend poleward

    b. isotherms in North America trend east and west or zonal, parallel to the equator

    c. the coldest region on the map is in central Canada

    d. isotherms trend equatorward in the continental interiors of the Northern Hemisphere

6. Our individual perception of temperature is termed

    a. sensible heat

    b. air temperature

    c. the heat index

    d. apparent temperature

7. Relative to future temperatures,

    a. humans cannot influence long-term temperature trends

    b. short-term changes appear to be out of our reach to influence

    c. a cooperative global network of weather monitoring among nations has yet to be established

    d. human society is causing short-term changes in global temperatures and temperature patterns

8. Evaporation

    a. tends to lower temperatures more over land as compared to over water bodies

    b. ends to moderate temperatures more over water bodies as compared to over land

    c. tends to increase the temperature over water

    d. affects land more than ocean surfaces

9. Sensible heat energy present in the atmosphere is expressed as air temperature.

    a. true

    b. false

10. Air temperature is usually measured with alcohol or mercury thermometers.

    a. true

    b. false

11. Both the Celsius and Fahrenheit temperature scales are in widespread general use worldwide.

    a. true

    b. false

12. Air temperatures, as reported by the National Weather Service or Canadian Meteorological Center, are always measured in full sunlight.

    a. true

    b. false

13. Wichita is more continental, whereas Trondheim is more marine in terms of climatic influences.

    a. true

    b. false

14. Surprisingly, the lowest natural temperature ever recorded on Earth was in the Southern Hemisphere.

    a. true

    b. false

15. The greatest annual range of temperatures on Earth occurs in north central Asia.

    a. true

    b. false

16. The temperature at which all motion in a substance stops is called 0° absolute temperature. Its equivalent in different temperature-measuring schemes is _____ Celsius (C), _____ Fahrenheit (F), and _____ Kelvin. The Fahrenheit scale places the melting point of ice at _____ °F (_____ °C, _____ K) and the boiling point of water at _____ °F (_____ °C, _____ K). The _____ is the only major country still using the Fahrenheit scale.

17. The coldest natural temperature ever recorded on Earth was (station, location, date, temperature reading):

_____

18. The warmest natural temperature ever recorded on Earth was (station, location, date, temperature reading):

_____

19. Try to find out the record high and record low temperatures for the city or town you are in as you work on this study guide. Perhaps your teacher or lab assistant can provide you with the information.

   City/Town (location of weather instruments): _____

   Minimum temperature: _____

   Maximum temperature: _____

   Average January temp.: _____

   Average July temp.: _____

20. Given the discussion of "Principal Temperature Controls" in Chapter 5, briefly analyze these high and low readings and average values for your city. What factors seem to be affecting temperature patterns in your town?

_____

_____

_____

6

ATMOSPHERIC AND OCEANIC CIRCULATIONS

Earth's atmospheric circulation is an important transfer mechanism for both energy and mass. In the process, the energy imbalance between equatorial surpluses and polar deficits is partly resolved, Earth's weather patterns are generated, and ocean currents are produced. Human-caused pollution also is spread worldwide by this circulation, far from its point of origin.

In this chapter we examine the dynamic circulation of Earth's atmosphere that carried Mount Pinatubo's debris worldwide and also carries the everyday ingredients oxygen, carbon dioxide, and water vapor around the globe. The wind-sculpted tree in the chapter opening photo brings home the power of atmospheric circulation and its potential for electrical generation.

The keys to this chapter are in several integrated figures: the portrayal of winds by the *Seasat* image (Figure 6.6), the three forces interacting to produce surface wind patterns (Figure 6.8), global barometric pressure on January and July maps (Figure 6.11), and the synthesis of concepts in the idealized illustration of global circulation in Figure 6.13.

OUTLINE HEADINGS AND KEY TERMS

The first-, second-, and third-order headings that divide Chapter 6 serve as an outline for your notes and studies. The key terms and concepts that appear **boldface** in the text are listed here under their appropriate heading in *bold italics*. All these highlighted terms appear in the text glossary. Note the check-off box (❏) so you can mark your progress as you master each concept. These terms should be in your reading notes or used to prepare note cards. The ⊛ icon indicates that there is an accompanying animation on the Student CD. The ✿ icon indicates that there is an accompanying satellite or notebook animation on the Student CD.

The outline headings and terms for Chapter 6:

Wind Essentials

Air Pressure and Its Measurement

❏ *mercury barometer*
❏ *aneroid barometer*

Wind: Description and Measurement

❏ *wind*
❏ *anemometer*
❏ *wind vane*

Global Winds

Driving Forces Within the Atmosphere

⊛ **Coriolis Force**

⊛ **Wind Pattern Development**

❏ *pressure gradient force*
❏ *Coriolis force*
❏ *friction force*

Pressure Gradient Force

❏ *isobar*

Coriolis Force

❏ *geostrophic winds*

Friction Force

❏ *anticyclone*
❏ *cyclone*

Atmospheric Patterns of Motion

⊛ **Global Wind Circulation, Hadley Cells**

✿ **Global Infrared Satellite Loop**

URLs listed in Chapter 6

Traditional Beaufort scale:
http://www.crh.noaa.gov/lot/webpage/beaufort/

Indian Ocean experiment:
http://www-indoex.ucsd.edu/ProjDescription.html

Multiyear Oscillations in Global Circulation:
http://www.ldeo.columbia.edu/NAO/
http://topex-www.jpl.nasa.gov/science/pdo.htm

Wind energy generation:
http://www.awea.org/
http://www.acga.org/programs/2001WindCCS/

KEY LEARNING CONCEPTS FOR CHAPTER 6

The following key learning concepts help guide your reading and comprehension efforts. The operative word is in *italics*. Use these carefully to guide your reading of the chapter and note that STEP 1 asks you to work with these concepts. These same learning concepts are used in organizing the summary and review section at the end of the chapter—grouping together definitions, a list of key terms, and review questions.

After reading the chapter and using this study guide, you should be able to:

- *Define* the concept of air pressure, and *describe* instruments used to measure air pressure.
- *Define* wind, and *describe* how wind is measured, how wind direction is determined, and how winds are named.
- *Explain* the four driving forces within the atmosphere—gravity, pressure gradient force, Coriolis force, and friction force—and *describe* the primary high- and low-pressure areas and principal winds.
- *Describe* upper-air circulation and its support role for surface systems, and *define* the jet streams.
- *Overview* several multiyear oscillations of air temperature, air pressure, and circulation in the Arctic, Atlantic, and Pacific oceans.
- *Explain* several types of local winds: land–sea breezes, mountain–valley breezes, katabatic winds, and the regional monsoons.
- *Discern* the basic pattern of Earth's major surface ocean currents and deep thermohaline circulation.

❋ STEP 1: Critical Thinking Process

Using your interest and learning, and the following questions as guidelines <u>only</u>, briefly discuss your experience with this chapter. In examining your learning you need not go through each of these questions in detail, simply provide an overview of your critical thinking process as it relates to some aspect of this chapter.

- What did you know about the learning concept before you began?
- Which information sources did you use in your learning (text, class, other)?
- Were you able to complete the action stated in the learning concept? What did you learn?
- Are there any aspects of the concept about which you want to know more?

Critical Thinking and Chapter 6: _____

_____

_____

_____

_____

❋ STEP 2: Global Winds Create Global Linkages and Effects

1. Examine Figure 6.1 (b) through (e), read the caption and the related section of the text. Answer the following completion items and questions.

(a) Describe the event depicted in the figure. _____

_____

(b) What remote-sensing instrument aboard which satellite made these images?

_____

(c) What is AOT and how does it appear on the images? _____

_____

(d) Describe the progression of aerosols between June 15 and August 21, 1991.

_____

_____

_____

(e) In Chapter 1, review the map in Figure 1.6. What were some of the effects of this eruption?

_____

_____

_____

❋ STEP 3: Air Pressure and Driving Forces

1. Locate a barometer either at the college or university you are attending, or perhaps at home. Or find a local weather broadcast on television or radio, a local cable channel, the Weather Channel, a local newspaper, the National Weather Service or Environment Canada that reliably presents barometric pressure information. For at least 3 days, record the air pressure each day at approximately the same time, if possible. See if you can detect a trend or the relationship between air pressure and other atmospheric phenomena. Spaces are provided here for you to record your observations in either millibars or inches. Note: you can do approximate conversions using the scale below Figure 6.3.

Air pressure observations

Day 1: _____ mb _____ in. Place: _____ Date: _____

Day 2: _____ mb _____ in. Place: _____ Date: _____

Day 3: _____ mb _____ in. Place: _____ Date: _____

2. Relative to air pressure:

(a) What is normal sea-level pressure expressed in mb and in.? _____

(b) What instruments are used to measure air pressure? _____

_____

(c) According to the text and Figure 6.3, the following records are mentioned for air pressure.

Earth's record low pressure: _____

U.S. record low pressure: _____

U.S. record high pressure: _____

Canada record low pressure: _____

Canada record high pressure: _____

Earth's record high pressure: _____

3. Using the Beaufort wind scale in Table 6.1, complete the following table.

Wind Speed			Beaufort Number	Wind Description	Observed Effects at Sea	Observed Effects on Land
kmph	mph	knot				
			1			
			4			
			7			
			9			
			11			

4. Label the 16 wind directions in the spaces provided on this wind compass.

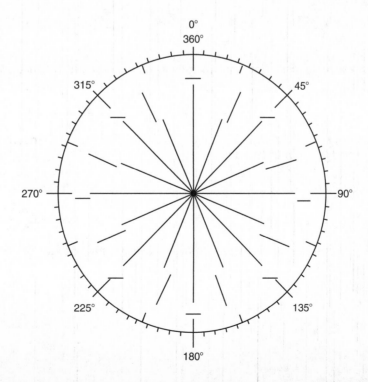

5. Is there an operating wind vane and anemometer on your campus? If so, where are they installed? Where are the instruments kept that register their measurements?

_____

_____

✳ STEP 4: Global Winds

1. An image from the *Seasat* satellite appears in Figure 6.6. Answer the following completion items and questions.

(a) Can you identify the northeast and southeast trade winds as they <u>converge</u> along the intertropical convergence zone north of the equator? If so, describe the location and pattern.

_____

_____

(b) Can you locate a large anticyclonic (clockwise) circulation system in the north Pacific Ocean (consult the small white arrows)? Describe the location and mention the direction of the wind flow. (See Figure 6.8c.)

_____

_____

(c) Can you locate a cyclonic (clockwise) circulation in the south Pacific Ocean? If so, describe the location and mention the direction of the wind flow. (See Figure 6.8c.)

_____

_____

(d) From the caption and text, describe how this image was made, when it was made, and where the image was obtained.

_____

2. Look ahead to the model of global winds in Figure 6.13 and make any comparisons of the patterns in the *Seasat* image with those portrayed in the model. _____

_____

3. Summarize the three driving forces within the atmosphere other than gravity. Briefly define and describe the influence of each:

(a) _____ : _____

_____

(b) _____ : _____

_____

(c) _____ : _____

_____

Note: Use Figure 6.8 to integrate these three driving force concepts. Each force is added in sequence with the resultant effects on winds illustrated. (a) Shows pressure gradient force only. (b) Adds the Coriolis deflection force to the pressure gradient force to produce winds characteristic of upper-troposphere geostrophic winds. (c) Summarizes all three forces by adding the friction force to the other two to produce characteristic surface winds.

4. Carefully examine the two maps in Figure 6.11. These portraits of January and July barometric pressure will help you understand the pattern of global winds. Compare and contrast specific regions on each map. For instance, take the North Pacific Ocean in January (dominated by sub-polar low pressure—the Aleutian low); then, in July, the same approximate area is dominated by the northern margins of an enormous Pacific high that dominates the West Coast.

_____

_____

_____

_____

_____

Compare Asia in January and the location of the ITCZ (dipping down over northern Australia), with July and an ITCZ that is shifted north of India. In the Southern Hemisphere midlatitudes, how would you characterize the pressure gradient? Do you think the winds produced are going to be strong? Consistent from season to season?

_____

_____

_____

_____

_____

5. Describe the migration of the subtropical high-pressure cells as they follow the subsolar point. How much do they migrate in degrees of latitude?

_____

_____

_____

_____

6. How do the western sides of the subtropical high-pressure cells differ from the eastern sides? How do these differences affect weather patterns on the continents?

_____

_____

_____

_____

✳ STEP 5: Wind Patterns

1. Complete Earth's primary high and low pressure areas from Table 6.2, "Four Hemispheric Pressure Areas."

Name	Cause	Location	Air temperature/ moisture

2. In the space provided, reproduce in a simple diagram the general atmospheric circulation depicted in Figure 6.13a and add appropriate labels. Begin by marking the North and South Poles, draw the equator, and make approximate latitudinal marks at 15°, 30°, 45°, 60°, and 75° to guide your sketch.

3. Can you identify the ITCZ on the *Galileo* remote-sensing image in Figure 6.12 (cloud patterns)?

Briefly describe it: _____

_____

Describe the subtropical high-pressure regions (clear skies) on this image: _____

_____

4. How many subpolar-low cyclonic systems (clockwise flow in the Southern Hemisphere) do you see in the *Galileo* remote-sensing image in Figure 6.15, (surrounding Antarctica)? Briefly discuss: _____

_____

_____

_____

5. Examine Figure 6.14, and note the clockwise winds surrounding the gyre in the North Atlantic Ocean. Using these concepts shift to the map of the North Pacific Ocean in News Report 6.3, Figure 6.3.1, and describe what is going on with a "message in a bottle and rubber duckies."

_____

_____

_____

6. Describe the effects of the jet stream on weather patterns in the midlatitudes. How do jet streams affect flight times (News Report 6.2)?

_____

_____

_____

_____

7. Describe the two phases of the North Atlantic Oscillation (NAO). How does each phase affect weather patterns in the eastern U.S. and Europe?

_____

_____

_____

_____

8. Describe the two phases of the Arctic Oscillation (AO). How is the AO related to the NAO?

_____

_____

_____

_____

✳ STEP 6: Wind Power

1. Assess the potential for wind-generated electricity as outlined in Focus Study 6.1. Is wind power economical? What appears to be the major stumbling block to development?

_____

_____

_____

_____

2. Briefly describe present wind installations (location and amount). Which countries are the leaders through 2003 (see Figure 6.1.2)? Which U.S. states have the greatest potential for wind power development?

_____

_____

_____

3. In California how are demand and production patterns related?

_____

_____

_____

4. Related to the power of the wind, how do the events described in News Report 6.3 help explain the circulation patterns in the North Pacific ocean?

_____

_____

_____

✳ STEP 7: Monsoons and Local Winds

1. Figure 6.21a and b present the Asian monsoonal patterns for January and July. On the following outline version of the illustration, __complete__ the labeling and __add__ directional arrows to indicate wind patterns as shown on the maps in the text.

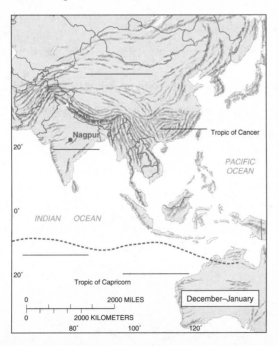
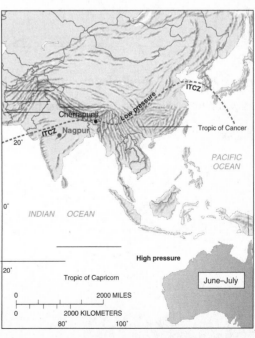

2. Characterize land-sea breezes (Figure 6.19): _____

_____

_____

3. Characterize mountain-valley breezes (Figure 6.20): _____

_____

_____

✳ STEP 8: Oceanic Currents

1. How is the Pacific gyre's circulation is related to Earth's air pressure systems and wind circulation patterns. Which pressure belts and wind patterns control the movement of the northern and southern portions of the gyre? _____

_____

_____

_____

2. What is the western intensification? List at least two examples of this phenomenon, as well as the amount of increased sea level height. _____

_____

_____

3. Describe the course of thermohaline circulation. Be specific about where the current is rising and sinking, as well as where it is absorbing and releasing heat energy. What are the possible effects of a slowing down or stopping of this circulation? Are there any signs that this is occurring? _____

_____

_____

_____

_____

❋ STEP 9: NetWork—Internet Connection

There are many Internet addresses (URLs) listed in this chapter of the *Geosystems* textbook. Go to any two URLs, or "Destination" links on the *Geosystems* Home Page, and briefly describe what you find.

1. _____ : _____

_____

2. _____ : _____

_____

SAMPLE SELF-TEST

(Answers appear at the end of the study guide.)

1. Normal sea level pressure has a value of

   a. 1013.2 millibars or 760 mm of mercury
   b. 28.50 inches of lead or dirt
   c. 32.01 inches of mercury
   d. 506.5 millibars

2. The horizontal motion of air relative to Earth's surface is generally considered

   a. barometric pressure
   b. wind
   c. convection flow
   d. an indicator of temperature

3. Which of the following describes the friction force?

   a. drives air from areas of higher to lower barometric pressure
   b. decreases with height above the surface
   c. causes apparent deflection of winds from a straight path
   d. drives air from areas of lower to higher barometric pressure

4. The combined effect of the Coriolis force and the pressure gradient force produces

   a. geostrophic winds
   b. any surface wind
   c. trade winds
   d. air flow moving directly between high and low pressure centers

5. Between 20° to 35° north and 20° to 35° south latitudes, you find

   a. the largest zone of water surpluses in the world
   b. strong westerly winds
   c. the world's arid and semi-arid desert regions and subtropical high pressure
   d. cyclonic systems of low pressure

6. The east side of subtropical high pressure cells (off continental west coasts) tend to be

   a. cool and moist
   b. warm, moist, and unstable
   c. dry, stable, and warm, with cooler ocean currents
   d. generally in the same position all year, not migrating with the high Sun

7. Land-sea breezes are caused by

   a. the fact that water heats and cools faster than land surfaces do
   b. cooler air flowing offshore (toward the ocean) in the afternoon
   c. onshore (toward the land) air flows develop in the afternoon as the land heats faster than the water surfaces
   d. the presence of mountain ranges

8. Monsoonal winds are

   a. regional wind systems that vary seasonally
   b. limited to only the Indian subcontinent
   c. a form of mountain-valley wind
   d. unrelated to the ITCZ position

9. An isoline of equal pressure plotted on a weather map is known as

   a. an isotherm
   b. an equilibrium line
   c. an isobar
   d. the thermal equator

10. Air pressure is unrelated to the temperature of the atmosphere.

    a. true
    b. false

11. Air pressure is measured with either a mercury or aneroid barometer.

    a. true
    b. false

12. Atmospheric circulation remains unrelated to the Limited Test Ban Treaty of 1963.

    a. true
    b. false

13. Wind is principally measured with a wind vane and an anemometer.

    a. true
    b. false

14. An area of high pressure circulates counterclockwise in the Northern Hemisphere and is called an anticyclone.

    a. true
    b. false

15. Winds flow from higher to lower pressure areas as a result of friction force.

    a. true
    b. false

16. Geostrophic winds are surface winds that form in response to the friction force.

    a. true
    b. false

17. A high pressure area is called an anticyclone, a low pressure area a cyclone.

   a. true
   b. false

18. The subtropical belt of high pressure is the place of the intertropical convergence zone.

   a. true
   b. false

19. The principal centers of low pressure in the Northern Hemisphere are the Aleutian and the Icelandic lows associated with the polar front, especially in winter.

   a. true
   b. false

20. Ocean currents play a relatively small role in regulating climate.

   a. true
   b. false

21. The European Union and several other countries are far outpacing the United States in the installation of wind power.

   a. true
   b. false

22. The western intensification refers to: _____

   _____

   _____

   _____

   _____

23. Characterize the circulation in the Pacific Ocean described in News Report 6.3, "A Message in a Bottle and Rubber Duckies."

   _____

   _____

   _____

24. The four forces that shape the speed and direction of wind are:

   _____; _____; _____; _____

PART TWO:

The Water, Weather, and Climate Systems

OVERVIEW—PART TWO

Part Two presents spatial aspects of hydrology, meteorology and weather, water resources, and climate. We begin with water itself—its origin, location, distribution, and properties. Earth is unique in its role as the water planet in our Solar System. Water vapor in the atmosphere stores and delivers vast amounts of energy to power the weather system. Conditions of stability and instability determine the potential for precipitation and weather activity.

The dynamics of daily weather phenomena include the interpretation of cloud forms, the interaction of air masses, and the occurrence of violent weather.

The specifics of the hydrologic cycle are explained through the water-balance concept, which is useful in understanding water-resource relationships, whether global, regional, or local. Important water resources include rivers, lakes, groundwater, and oceans. The spatial implications over time of this water-weather system lead to the final topic in Part Two, world climate patterns, and significantly, the present status of the ongoing global climate change. (A systems flow-schematic of Part 2 organization is in Figure 1.7b.)

Name: _____   Class Section: _____

Date: _____   Score/Grade: _____

7

WATER AND ATMOSPHERIC MOISTURE

Water is the essential medium of our daily lives and a principal compound in nature. Water covers 71% of Earth (by area), and within the solar system occurs in such significant quantities only on our planet. Water constitutes nearly 70% of our bodies by weight and is the major ingredient in plants, animals, and our food. A human being can survive 50 to 60 days without food, but only 2 or 3 days without water. The water we use must be adequate in quantity as well as quality for its many tasks, from personal hygiene to vast national water projects. Indeed water occupies the place between land and sky, mediating energy and shaping both the lithosphere and the atmosphere.

This chapter examines the dynamics of water, water vapor, and its role in atmospheric stability and instability, water and water vapor in the atmosphere in the form of clouds and fog—the essentials of weather. The clouds that form as air becomes water-saturated are more than whimsical, beautiful configurations; they are important indicators of overall atmospheric conditions.

OUTLINE HEADINGS AND KEY TERMS

The first-, second-, and third-order headings that divide Chapter 7 serve as an outline for your notes and studies. The key terms and concepts that appear **boldface** in the text are listed here under their appropriate heading in ***bold italics***. All these highlighted terms appear in the text glossary. Note the check-off box (❑) so you can mark your progress as you master each concept. These terms should be in your reading notes or used to prepare note cards. The ⊙ icon indicates that there is an accompanying animation on the Student CD.

The outline headings and key terms for Chapter 7:

Water on Earth

⊙ Earth's Water and the Hydrologic Cycle
- ❑ *outgassing*

Worldwide Equilibrium
- ❑ *eustasy*
- ❑ *glacio-eustatic*

Distribution of Earth's Water Today

Unique Properties of Water

Heat Properties
⊙ Water Phase Changes
- ❑ *phase change*
- ❑ *sublimation*

Ice, the Solid Phase
Water, the Liquid Phase
- ❑ *latent heat*

Water Vapor, the Gas Phase
- ❑ *latent heat of vaporization*
- ❑ *latent heat of condensation*

Heat Properties of Water in Nature
- ❑ *latent heat of sublimation*

Humidity
- ❑ *humidity*

Relative Humidity
- ❑ *relative humidity*

Saturation
- ❑ *saturated*
- ❑ *dew-point temperature*

Daily and Seasonal Relative Humidity Patterns

Expressions of Relative Humidity
Vapor Pressure
- ❑ *vapor pressure*

Specific Humidity
- ❑ *specific humidity*

Instruments for Measuring Humidity
- ❑ *hair hygrometer*
- ❑ *sling psychrometer*

Atmospheric Stability

Atmospheric Stability
- ❑ *stability*

Adiabatic Processes
- ❑ *adiabatic*

Dry Adiabatic Rate
- ❑ *dry adiabatic rate (DAR)*

Moist Adiabatic Rate
- ❑ *moist adiabatic rate (MAR)*

Stable and Unstable Atmospheric Conditions

Clouds and Fog
- ❑ *cloud*

Cloud Formation Processes
- ❑ *moisture droplet*
- ❑ *cloud-condensation nuclei*

Cloud Types and Identification
- ❑ *stratus*
- ❑ *nimbostratus*
- ❑ *cumulus*
- ❑ *stratocumulus*
- ❑ *altocumulus*
- ❑ *cirrus*
- ❑ *cumulonimbus*

Fog
- ❑ *fog*

Advection Fog
- ❑ *advection fog*
- ❑ *evaporation fog*
- ❑ *upslope fog*
- ❑ *valley fog*

Radiation Fog
- ❑ *radiation fog*

SUMMARY AND REVIEW

News Reports

News Report 7.1: Breaking Roads and Pipes and Sinking Ships
News Report 7.2: Harvesting Fog

URLs listed in Chapter 7

Fog harvesting for water:
http://www.idrc.ca/nayudamma/fogcatc_72e.html

Ice and snowflakes:
http://www.its.caltech.edu/~atomic/snowcrystals/
http://www.lpsi.barc.usda.gov/emusnow/

Related URLs for topics in this chapter are found in Chapters 8, 9, 14, and 16.

KEY LEARNING CONCEPTS FOR CHAPTER 7

The following key learning concepts help guide your reading and comprehension efforts. The operative word is in *italics*. Use these carefully to guide your reading of the chapter and note that STEP 1 asks you to work with these concepts. These same learning concepts are used in organizing the summary and review section at the end of the chapter—grouping together definitions, a list of key terms, and review questions.

After reading the chapter and using this study guide, you should be able to:

- *Describe* the origin of Earth's waters, *define* the quantity of water that exists today, and *list* the locations of Earth's freshwater supply.
- *Describe* the heat properties of water, and *identify* the traits of its three phases: solid, liquid, and gas.
- *Define* humidity and the expressions of the relative humidity concept, and *explain* dew-point temperature and saturated conditions in the atmosphere.
- *Define* atmospheric stability, and *relate* it to a parcel of air that is ascending or descending.
- *Illustrate* three atmospheric conditions—unstable, conditionally unstable, and stable—with a simple graph that relates the environmental lapse rate to the dry adiabatic rate (DAR) and moist adiabatic rate (MAR).
- *Identify* the requirements for cloud formation, and *explain* the major cloud classes and types, including fog.

❋ **STEP 1: Critical Thinking Process**

Using your interest and learning, and the following questions as guidelines <u>only</u>, briefly discuss your experience with this chapter. In examining your learning you need not go through each of these questions in detail, simply provide an overview of your critical thinking process as it relates to some aspect of this chapter.

- What did you know about the learning concept before you began?
- Which information sources did you use in your learning (text, class, other)?
- Were you able to complete the action stated in the learning concept? What did you learn?
- Are there any aspects of the concept about which you want to know more?

Critical Thinking and Chapter 7: _____

_____

_____

_____

✳ STEP 2: Water Essentials

1. If deprived of food and water, describe how long a human can survive without each. _____

2. How much water constitutes Earth's hydrosphere (km^3 and mi^3)? _____

_____

3. What does the photograph in Figure 7.1 demonstrate? Explain. _____

_____

_____

_____

4. The seven largest lakes in the world in terms of volume represent _____% of Earth's freshwater (Table 7.1 inset). In one lake, _____, some _____% of Earth's freshwater is found. (Hint: divide the largest lake by the total volume in freshwater lakes to get the percent.)

5. The three largest lakes in the world in terms of volume are

      <u>Lake</u>   <u>Volume $km^3(mi^3)$</u>

      a) _____   _____

      b) _____   _____

      c) _____   _____

6. Relative to the four oceans:

      <u>Ocean</u>   <u>% of Earth's ocean area</u>

      a) _____   _____

      b) _____   _____

      c) _____   _____

      d) _____   _____

7. What <u>percentage</u> of <u>freshwater</u> is represented by

    Ice sheets and glaciers? _____   Rivers and streams? _____

    Freshwater lakes? _____   All groundwater? _____

    Saline lakes and inland seas? _____   Soil moisture storage? _____

    Atmosphere? _____

✳ STEP 3: Water Properties

1. *Name* the term that describes the following phase changes and *list* the latent heat energy (in calories) that is either absorbed or released for one gram making the change in state. (See Figures 7.4 and 7.6.)

(a) solid to liquid (at 0°C): _____

(b) liquid to vapor (at 20°C): _____

(c) liquid to solid (at 0°C): _____

(d) vapor to liquid (at 100°C): _____

2. The three illustrations in Figure 7.5 demonstrate the properties of ice and the hydrogen bonding between water molecules. Explain. _____

_____

_____

3. So what's with breaking roads and pipes and sinking ships (News Report 7.1)? _____

_____

_____

4. Given the quantity of latent heat energy needed to evaporate water and that is liberated when water vapor condenses, what does this tell you about the power that must be in clouds, given that a small, puffy cumulus cloud weighs between 500 and 1000 tons? _____

_____

_____

✳ STEP 4: Humidity and Stability

1. Describe two expressions of relative humidity.

(a)_____

(b)_____

2. What happens in a parcel of air when the *dew-point temperature* and the *air temperature* are the same? Begin by defining dew-point temperature.

_____

_____

3. What happens on the outside of a glass when you fill it with ice water, as shown in the illustration in Figure 7.9a? _____

_____

_____

4. Describe the operation of two instruments used to measure relative humidity (Figure 7.14).

(a) _____ : _____

_____

(b) _____ : _____

_____

5. A parcel of air is at 20°C (68°F) and has a saturation vapor pressure of 24 mb. If the water vapor content actually present is exerting a vapor pressure of only 12 mb in 20°C air, the relative humidity is _____. What would the approximate relative humidity be if the parcel of air increased in temperature to 30°C (86°F), and the new saturation vapor pressure of 40 mb (12 mb ÷ 40 mb)? _____
What would the approximate relative humidity be if the parcel of air decreased in temperature to 10°C (50°F)?_____
Use Figure 7.12 in preparing your answers.

6. Using the graph in Figure 7.11, *compare* and *contrast* the mean relative humidity and air temperature for a series of typical days.

(a) Relative humidity: _____

_____

_____

_____

(b) Air temperature: _____

_____

_____

(c) What is the relationship between these two variables shown on the graph?

_____

_____

_____

7. Using the principles described in the text and Figure 7.15, describe what you see in the photograph in Figure 7.16. How are these principles demonstrated in the photo?

_____

_____

_____

8. Figure 7.17 illustrates adiabatic principles. Explain what is occurring in each diagram:

Figure 7.17a _____

_____;

Figure 7.17b _____

_____

9. Plot data lines that denote the *dry adiabatic rate* (*DAR*) and *moist adiabatic* (*MAR*) rates (see Figure 7.19). Label each line. Label portions of the graph that represent atmospheric conditions of: **a)** <u>unstable</u>, **b)** <u>conditionally unstable</u>, and **c)** <u>stable</u>—given different values for the *environmental lapse rate*.

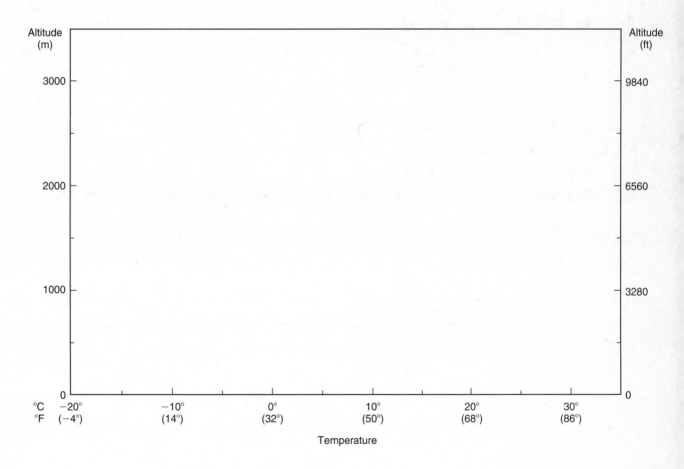

10. Describe atmospheric conditions of stability (unstable and stable) portrayed in the following two graphs (Figure 7.20).

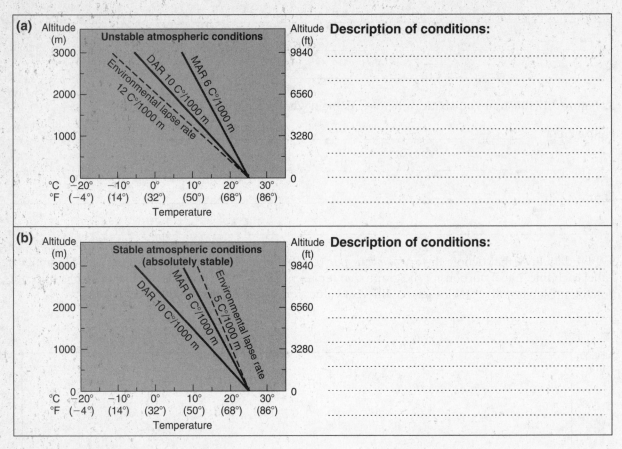

(a)

Altitude (m) / Altitude (ft)

Unstable atmospheric conditions

Environmental lapse rate 12 C°/1000 m

DAR 10 C°/1000 m

MAR 6 C°/1000 m

3000 / 9840
2000 / 6560
1000 / 3280
0 / 0

°C −20° −10° 0° 10° 20° 30°
°F (−4°) (14°) (32°) (50°) (68°) (86°)

Temperature

Description of conditions:
....
....
....
....
....
....

(b)

Altitude (m) / Altitude (ft)

Stable atmospheric conditions (absolutely stable)

Environmental lapse rate 5 C°/1000 m

MAR 6 C°/1000 m

DAR 10 C°/1000 m

3000 / 9840
2000 / 6560
1000 / 3280
0 / 0

°C −20° −10° 0° 10° 20° 30°
°F (−4°) (14°) (32°) (50°) (68°) (86°)

Temperature

Description of conditions:
....
....
....
....
....
....

✳ STEP 5: Clouds and Fog

1. What is a cloud? How do moisture droplets form? Are clouds made of rain drops (Figures 7.21, 7.22)?

_____

_____

_____

2. Using Figure 7.23, Table 7.2, and Figure 7.24, describe the physical characteristics of the cloud types listed below:

Cloud type	Description
a. stratus:	_____
	_____
b. cumulus:	_____
	_____
c. nimbostratus:	_____
	_____

d. cumulonimbus: _____

_____

e. altostratus: _____

_____

f. cirrus: _____

_____

g. fog: _____

_____

3. Name several areas in the United States and Canada that experience the highest annual number of days with heavy fog (see the map and caption in Figure 7.30).

_____

_____

✳ STEP 6: NetWork—Internet Connection

There are many Internet addresses (URLs) listed in this chapter of the *Geosystems* textbook. Go to any two URLs, or "Destination" links on the *Geosystems* Home Page, and briefly describe what you find.

1. _____ : _____

_____

2. _____ : _____

_____

SAMPLE SELF-TEST

(Answers appear at the end of the study guide.)

1. Water covers approximately what percentage of Earth's surface?

    a. 50%

    b. 100%

    c. 71%

    d. 83%

    e. less than 50%

2. The present quantity of water on Earth was achieved approximately _____ years ago, according to the text.

    a. one million years ago

    b. one billion years ago

    c. two billion years ago

    d. since the last ice age

    e. two million years ago

3. Specifically, glacio-eustatic factors refer to

    a. worldwide changes in land masses

    b. a steady-state equilibrium in the water system

    c. changes in water location related to the increase or decrease in quantities of ice

    d. worldwide changes in sea level

4. Water has unusual heat properties, partially related to hydrogen bonding, compared with other compounds. If you were to take one gram of ice at 0°C and raise it to one gram of water vapor at 100°C, how many calories would you need to add in terms of latent and sensible heat?

    a. 540 calories

    b. 80 calories

    c. 320,000 calories per square centimeter

    d. 720 calories

5. The major portion of freshwater today is located in

    a. all sub-surface water

    b. groundwater resources

    c. ice sheets and glaciers

    d. the major rivers and lakes and atmospheric moisture

    e. the oceans

6. Relative humidity refers to

    a. the amount of water vapor in the air compared with normal levels

    b. the amount of moisture in the air relative to your own sensible feelings

    c. the actual humidity in the air, or the absolute humidity

    d. the amount of water vapor in the air at a given temperature and pressure, expressed as a percentage of the moisture capacity of the air

    e. an unused concept when it comes to weather topics

7. Water molecules bind tightly to one another. This is a result of

    a. hydrogen bonding

    b. covalent bonding

    c. atomic friction

    d. molecular hold

8. What is the heat energy involved in the change of state, or phase, in water?

    a. mechanical heat

    b. sensible heat

    c. fusion heat

    d. latent heat

9. Assume that a warm air bubble, or parcel, near Earth's surface, that has a temperature of 25°C, begins to rise. Assume that the parcel of air contains 12 mb vapor pressure. At what altitude will the lifting mass of air become saturated? (Use the saturation vapor pressure graph in Figure 7.12, and a DAR of 1 C° per 100 m.)

    a. 400 m

    b. 1000 m

    c. 1500 m (15 C° of cooling to 10°C)

    d. it does not reach the dew-point temperature in this example

10. The dry adiabatic rate (DAR) is

    a. 6 C° per 1000 m (3.3 F° per 1000 ft)

    b. the rate used in lifting a saturated parcel of air

    c. the environmental lapse rate

    d. 10 C° per 1000 m (5.5 F° per 1000 ft)

    e. only used in the desert

11. Which of the following are <u>correctly</u> matched?

    a. flat or layered clouds = cumulus
    b. puffy or globular clouds = cirroform
    c. puffy or globular clouds = cumuliform
    d. wispy clouds = water droplets

12. Which of the following is a middle-level cloud type

    a. cirrostratus
    b. stratocumulus
    c. cumulonimbus
    d. altostratus

13. A cirrostratus cloud is associated with

    a. patches of cotton balls, dappled, or arranged in lines
    b. Sun's outline just visible
    c. veil of ice crystals causing a halo around the Moon
    d. sharply outlined, billowy

14. Which cloud type is specifically a good indicator of an arriving storm, say within the next 24 hours?

    a. fog
    b. cumulus
    c. stratocumulus
    d. cirrus

15. A thunderstorm is associated with

    a. cirrostratus
    b. stratocumulus
    c. cumulonimbus
    d. nimbostratus

16. The prefix *nimbo-* and the suffix *-nimbus*

    a. mean that clouds are generally forming
    b. designates clouds that occur in the middle altitudes (2000–6000 m)
    c. mean that clouds are generally dissipating
    d. mean that the clouds are producing precipitation

17. Earth, like the other planets in the solar system, possesses large quantities of water.

    a. true
    b. false

18. Glacio-eustatic factors specifically relate to changes in sea level caused by actual physical changes in the elevation of landmasses.

    a. true
    b. false

19. The water vapor content of the air is termed humidity.

    a. true
    b. false

20. Air is saturated when the dew-point temperature and the air temperature coincide.

    a. true
    b. false

21. That portion of the total air pressure that is made up of water vapor molecules is termed specific humidity.

    a. true
    b. false

22. Relative humidity is a <u>direct</u> measure of the water vapor content of the air.

   a. true
   b. false

23. The phase change of water to ice is called freezing.

   a. true
   b. false

24. The energy involved in the phase changes of water is called latent heat.

   a. true
   b. false

25. Clouds are initially composed of raindrops.

   a. true
   b. false

26. The _____ uses the principle that human hair changes as much as 4% in length between 0 and 100% relative humidity. The _____ has two thermometers mounted side-by-side on a metal holder. One is called the dry-bulb thermometer; it simply records the ambient (surrounding) air temperature. The other thermometer is called the wet-bulb thermometer. That portion of total air pressure that is made up of water vapor molecules is termed _____ and is expressed in millibars (mb). _____ refers to the mass of water vapor (in grams) per mass of air (in kilograms) at any specified temperature.

27. An air parcel is considered <u>unstable</u> when

   a. it remains in its initial position
   b. it continues to rise until it reaches an altitude where the surrounding air has a similar density
   c. it resists displacement upward
   d. it ceases to ascend and begins to descend

28. Which cloud type dominates weather in the tropical rain forest region?

   a. cirrus
   b. altostratus
   c. cumulonimbus
   d. nimbostratus

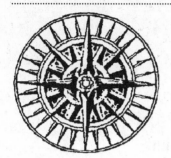

8

WEATHER

We begin our study of weather with a discussion of the huge air masses that move across North America. We observe powerful lifting mechanisms in the atmosphere, examine cyclonic systems, and conclude with a portrait of the violent and dramatic weather that occurs in the atmosphere. Temperature, air pressure, relative humidity, wind speed and direction, daylength, and Sun angle are important measurable elements that contribute to the weather.

Temperature, air pressure, relative humidity, wind speed and direction, daylength, and Sun angle are important measurable elements that contribute to the weather. For the weather forecast, we turn to the National Weather Service in the United States (*http://www.nws.noaa.gov*) or to the Canadian Meteorological Centre, a branch of the Meteorological Service of Canada (MSC), (*http://www.cmc.ec.gc.ca/*), to see current satellite images and to hear weather analysis. Internationally, the World Meteorological Organization coordinates weather information (see *http://www.wmo.ch/*). Many sources of weather information and related topics are found in the "Destinations" section for this chapter on the *Geosystems* Home Page.

OUTLINE HEADINGS AND KEY TERMS

The first-, second-, and third-order headings that divide Chapter 8 serve as an outline. The key terms and concepts that appear **boldface** in the text are listed here under their appropriate heading in ***bold italics***. All these highlighted terms appear in the text glossary. Note the check-off box (❑) so you can mark your progress as you master each concept. These terms should be in your reading notes or used to prepare note cards. The ✪ icon indicates that there is an accompanying animation on the Student CD. The ✺ icon indicates that there

is an accompanying satellite or notebook animation on the Student CD.

The outline headings and terms for Chapter 8:

Weather Essentials

    ❑ *weather*
    ❑ *meteorology*

Air Masses

    ❑ *air mass*

Air Masses Affecting North America

Air Mass Modification

Atmospheric Lifting Mechanisms

✺ **Convectional Heating and Tornado in Florida Satellite Loop**

✪ **Cold and Warm Fronts**

Convergent Lifting

    ❑ *convergent lifting*

Convectional Lifting

    ❑ *convectional lifting*

Orographic Lifting

    ❑ *orographic lifting*
    ❑ *chinook winds*
    ❑ *rain shadow*

Frontal Lifting (Cold and Warm Fronts)

    ❑ *cold front*
    ❑ *warm front*

Cold Front

    ❑ *squall line*

Warm Front

SUMMARY AND REVIEW

News Reports, Focus Study, and Career Link

News Report 8.1: Mountains Set Precipitation Records

News Report 8.2: May 2003 Tornado Outbreak in Tornado Alley

Focus Study 8.1: Forecasting Atlantic Hurricanes

Career Link 8.1: Tracy Smith, Research Meteorologist

URLs listed in Chapter 8

National Weather Service:
http://www.nws.noaa.gov

Canadian Meteorological Centre:
http://www.msc-smc.ec.gc.ca/cmc/index_e.html

World Meteorological Organization:
http://www.wmo.ch/

Weather and severe storm info:
http://www.tornadoproject.com/fscale/fscale.htm
http://www.stormresearch.com/vortex/
http://www.nssl.noaa.gov
http://thunder.msfc.nasa.gov/lis/
http://www.joss.ucar.edu/vortex/
http://www.spc.noaa.gov/index.shtml
http://www.spc.noaa.gov/misc/AbtDerechos/derechofacts
    #definition
http://www.osei.noaa.gov/Events/Severe/US_Midwest/
    2003/SVRusMW125_G12.avi

Forecast Systems Laboratory, Boulder:
http://www.fsl.noaa.gov/

National Hurricane Center:
http://www.nhc.noaa.gov/

Hurricane forecasting:
http://typhoon.atmos.colostate.edu/forecasts/

All weather satellites:
https://www.npmoc.navy.mil/jtwc.html

KEY LEARNING CONCEPTS FOR CHAPTER 8

The following key learning concepts help guide your reading and comprehension efforts. The operative word is in *italics*. Use these carefully to guide your reading of the chapter and note that STEP 1 asks you to work with these concepts. These same learning concepts are used in organizing the summary and review section at the end of the chapter—grouping together definitions, a list of key terms, and review questions.

After reading the chapter and using this study guide, you should be able to:

- *Describe* air masses that affect North America, and *relate* their qualities to source regions.
- *Identify* types of atmospheric lifting mechanisms, and *describe* four principal examples.
- *Analyze* the pattern of orographic precipitation, and *describe* the link between this pattern and global topography.
- *Describe* the life cycle of a midlatitude cyclonic storm system, and *relate* this to its portrayal on weather maps.
- *List* the measurable elements that contribute to modern weather forecasting, and *describe* the technology and methods employed.
- *Analyze* various types of violent weather and the characteristics of each, and *review* several examples of each from the text.

✳ STEP 1: Critical Thinking Process

Using your interest and learning, and the following questions as guidelines <u>only</u>, briefly discuss your experience with this chapter. In examining your learning you need not go through each of these questions in detail, simply provide an overview of your critical thinking process as it relates to some aspect of this chapter.

- What did you know about the learning concept before you began?
- Which information sources did you use in your learning (text, class, other)?
- Were you able to complete the action stated in the learning concept? What did you learn?
- Are there any aspects of the concept about which you want to know more?

Critical Thinking and Chapter 8: _____

_____

_____

_____

_____

1. In your own words, define:

(a) weather _____

_____

_____

(b) meteorology:_____

_____

_____

(c) What is pictured in Figure 8.1? How many are there in the U.S.? Canada? _____

_____

_____

_____

STEP 2: Air Masses

1. Examine the map in Figure 8.4. Which air masses are creating this precipitation pattern?

_____

_____

2. Explain how the average annual snowfall patterns are produced as shown in Figure 8.5, based on air mass modification. _____

_____

_____

_____

3. Using the two maps in Figure 8.2, identify and add descriptive labels for the principal air masses that influence North America.

Winter pattern

Summer pattern

(a)

(b)

SST = Sea-surface temperature (°C)

✳ STEP 3: Atmospheric Lifting Mechanisms

1. List the four lifting mechanisms covered in the text and give a brief description of each (Figure 8.6).

(a) _____ : _____ ;

_____

(b) _____ : _____ ;

_____

(c) _____ : _____ ;

_____

(d) _____ : _____

_____

2. Figure 8.7 in the text presents conditions that lead to *convectional lifting*, condensation, and active cloud formation. Consider a parcel of air at Earth's surface to be at 25°C with 8 g/kg specific humidity (water vapor content).

Air at 25°C can hold 20 g/kg according to the graph in Figure 7.13. Since the air is holding 8 g/kg, with a maximum specific humidity of 20 g/kg, the air parcel is at 40% relative humidity at ground level. (8 g/kg [content] ÷ 20 g/kg [capacity] = 0.4 × 100 = 40%)

Assume that the environmental lapse rate of 12C° is in effect in the air surrounding the parcel. Using the maximum specific humidity graph in Figure 7.13, determine the dew-point temperature for this parcel of air—or where it reaches 100% relative humidity. The dew-point temperature occurs at the *lifting condensation level*, visible in the sky as the flat bottom of the clouds. Label all these components from Figure 8.7 on the following graph.

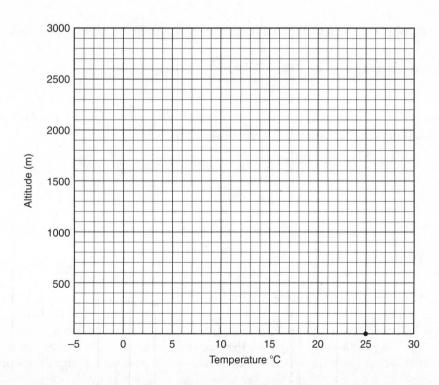

3. Questions and analysis of the graph figure.

(a) Label the figure with the information given in the first paragraph and place your answers to the following questions on the figure as well.

(b) What is the dew-point temperature of the lifting air parcel? _____

Show your work here: _____

(c) At what altitude does the lifting condensation level occur using the dry adiabatic rate of 1°C per 100 m? (The dew-point temperature is reached in the rising parcel of air and produces 100% relative humidity.)

_____ m; _____ ft.

Show your work here: _____

(d) Any further lifting of the parcel of air above this condensation level will produce cooling. Which adiabatic rate is used within a saturated parcel of air?

_____

4. Describe the conditions that led to the convectional activity over Florida as shown in the photo in Figure 8.8.

_____

_____

5. (a) After reading News Report 8.1, describe the place that is the wettest *average annual* precipitation on Earth and the location and physical characteristics that produce this total. _____

_____

(b) Describe the wettest place on Earth that holds the record for a *single year* and the location and physical characteristics that produce this total. _____

_____

6. Assess and describe the precipitation pattern in the State of Washington. Use a pair of windward and leeward stations to illustrate your explanation (Figure 8.10).

_____

_____

✳ STEP 4: A Weather Map

Weather maps used for showing atmospheric conditions at a specific time and place are known as *synoptic maps*. The daily weather map is a key analytical tool for meteorologists. An idealized passage of a midlatitude cyclone is shown in Figure 8.14, with model diagrams of each stage. Figure 8.14 also shows the common symbols and notations used on weather maps. Figure 8.15 in the text presents a portion of a weather map for April 20, 2000, along with a matching satellite image. Read the text before proceeding with the following.

Idealized Weather Map Analysis

Analyze the idealized weather map below to *determine* general weather conditions at the stations noted. Using the weather map symbols listed earlier in this lab exercise, *place* a weather symbol and information at each of the six city locations noted. Include: wind direction (if any), an estimate of wind speed, air pressure, air temperature, dew-point temperature, state of the sky, and a guess at weather type (if any)—all approximate, generalized estimates.

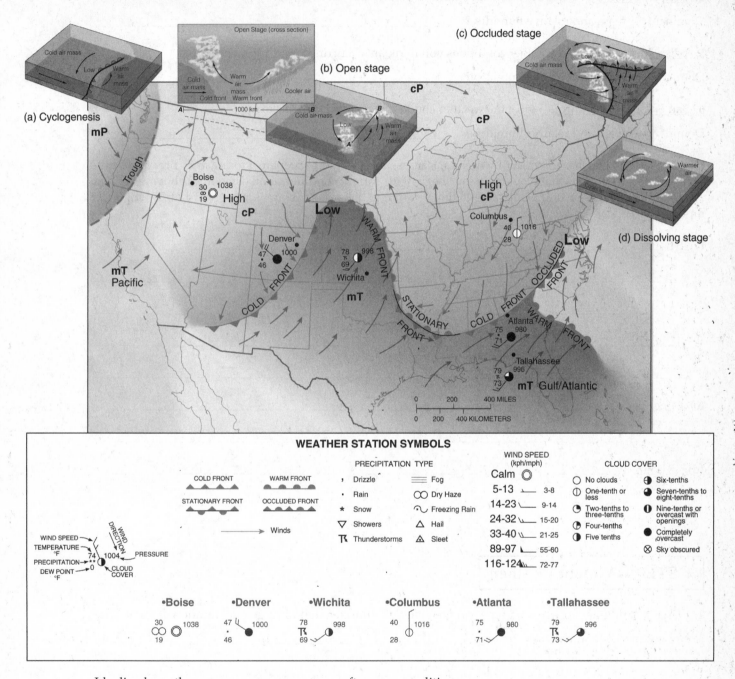

Idealized weather map—assume summer, afternoon conditions

Using the weather symbols presented earlier, complete the following for six cities.

1. Tallahassee, Florida—label the weather conditions you think are occurring.

Dominant air mass and relative humidity? _____

2. Columbus, Ohio—label the weather conditions you think are occurring.

Dominant air mass and relative humidity? _____

3. Wichita, Kansas—label the weather conditions you think are occurring.

Dominant air mass and relative humidity? _____

4. Denver, Colorado—label the weather conditions you think are occurring.

Dominant air mass and relative humidity? _____

5. Boise, Idaho—label the weather conditions you think are occurring.

Dominant air mass and relative humidity? _____

6. San Francisco, California—label the weather conditions you think are occurring.

Dominant air mass and relative humidity? _____

7. If you were a weather forecaster in Wichita, Kansas, and needed to prepare a forecast for broadcast covering the next 24 hours, what would you say? Begin with current conditions you listed in #3 above, and then progress through the next 6, 12, 24, and 48 hours. Assume the system is moving eastward at 25 kmph (15 mph), note the scale on the map.

(a) 6 hours: _____

_____

_____

(b) 12 hours: _____

_____

(c) 24 hours: _____

_____

(d) 48 hours: _____

_____

✳ STEP 5: Violent Weather

1. Describe a thunderstorm (cloud type and precipitation, thunder, lightning, general location of occurrence in North America, and hail potential). _____

_____

_____

_____

2. Relative to the satellite image in Figure 8.20: What is the satellite platform? What does the image show in terms of seasonal change in strikes? According to the text, what percentage of lightning strikes occur over land? _____

_____

_____

_____

_____

3. What are derechos and how are they related to thunderstorms? _____

_____

_____

4. Using the map in Figure 8.25, complete the following relative to tornado activity.

(a) Based on the average annual tornado incidence map in Figure 8.25a, which state has the greatest spatial intensity per 26,000 km^2? _____ Why do you think this is so?

_____

_____

(b) Using this map, describe tornado alley. _____

(c) In Figure 8.25b, monthly frequencies of tornadoes are reported. What are the top three months? _____, _____, _____ Why do you think these months top the year? _____

_____

_____

5. Relative to tornado records and annual averages, what has been occurring in the United States since 1990? _____

_____

_____

6. Compare Figure 8.19, showing thunderstorm patterns, with Figure 8.25, showing tornado patterns. How are they similar and how are they different? Why do you think this is? _____

_____

_____

_____

7. From Focus Study 8.1, text and map, overview the 1995 Atlantic hurricane season.

_____

_____

_____

8. In terms of hazard perception and damage, please assess the role of coastal settlement, construction, zoning, and the occurrence of hurricanes since Andrew in 1992. What is your assessment? What is needed?

_____

_____

_____

9. As global temperatures rise, more energy is available for evaporation. What effect would you expect this to have on the numbers and strengths of tornados and hurricanes? Have there been signs of unusual hurricane activity? Has the number of tornados and hurricanes increased over time? Have the costs of storm damage also increased? If so, by how much?

_____

_____

_____

_____

If you have access to the Internet be sure to check our _Geosystems Home Page_ for links to hurricane information and the present accuracy of the forecast for the Atlantic hurricane season given in the focus study. Our URL (Internet address) is: _http://www.prenhall.com/christopherson_. For Dr. William Gray's forecast method and latest Atlantic forecast see _http://typhoon.atmos.colostate.edu/forecasts/_.

10. Summarize Dr. Gray's latest forecast as given on his home page. Note that the summary is updated four

times a year. _____

_____

_____

✳ STEP 6: NetWork—Internet Connection

There are many Internet addresses (URLs) listed in this chapter of the _Geosystems_ textbook. Go to any two URLs, or "Destination" links on the _Geosystems_ Home Page, and briefly describe what you find.

1. _____ : _____

_____

2. _____ : _____

_____

SAMPLE SELF-TEST

(Answers at the end of the study guide.)

1. Which of the following weather conditions best describes your area today?

    a. dominance over the area by a high pressure cell
    b. a modified cP air mass
    c. the passage of a midlatitude wave cyclone
    d. local heating, creating high clouds and thundershowers
    e. a hurricane

2. Maritime tropical Pacific (mT) air and maritime tropical Gulf and Atlantic (mT) air

    a. are different from one another, since they occur over cool and warm ocean surfaces, respectively
    b. are identical in strength, since both carry the same mT destination
    c. are both usually present during the coldest weather conditions in the east and Midwest
    d. are not related to weather

3. The wettest average annual place on Earth is located

    a. in the United States
    b. in the Amazon in Brazil
    c. on the slopes of the Himalayas in India
    d. at the Monsoons in Southeast Asia

4. With respect to the three main lifting (cooling) mechanisms—local heating, orographic, and frontal—which of the following is <u>correct</u>?

    a. the place with the wettest average annual rainfall on Earth is most closely related to local heating and frontal activity
    b. a single convectional storm triggered by local heating affects large geographical regions
    c. adiabatic concepts apply only to local heating
    d. we do not get all three mechanisms within the United States
    e. given the necessary physical requirements, orographic precipitation is usually the most consistent type of the three

5. Lake-effect snowfall occurs

    a. on the leeward shores of the Great Lakes, since air masses become humidified over the lakes.
    b. over the Great Lakes
    c. most often along the windward shores of the lakes
    d. involving mT air masses

6. The wetter intercepting mountain slope is termed _____, as opposed to the drier far-side mountain slope, which is termed _____.

    a. westside; eastside
    b. leeward; windward
    c. wetside; dryside
    d. windward; leeward

7. Storm tracks across the United States and Canada generally

    a. shift with the high Sun—more southern storm tracks in winter, more northern storm tracks in summer
    b. move east to west
    c. move north to south
    d. do not change with the seasons at all

8. Tornado development is associated with
   a. warm fronts
   b. mesocyclone circulation and cold fronts
   c. continental tropical air masses
   d. stable air masses
   e. winter conditions in the Midwest

9. Summer afternoon thundershowers in the southeastern United States are more than likely a result of
   a. convectional lifting
   b. orographic lifting
   c. frontal lifting
   d. subtropical high pressure disturbance
   e. poor air quality

10. The world's rainfall records are associated with orographic precipitation.
    a. true
    b. false

11. A cold front is characterized by drizzly showers of long duration.
    a. true
    b. false

12. Cyclogenesis refers to the birth or strengthening of a wave cyclone.
    a. true
    b. false

13. Atmospheric <u>pressure</u> is portrayed on the daily weather map with a pattern of <u>isotherms</u>.
    a. true
    b. false

14. Thunder is produced by the sound of rapidly expanding air heated by lightning.
    a. true
    b. false

15. Typhoons and hurricanes are significantly different types of storms in terms of physical structure.
    a. true
    b. false

16. A tropical storm becomes a hurricane when its winds exceed 65 knots (74 mph).
    a. true
    b. false

17. Relative to orographic precipitation, the wetter intercepting slope is termed the _____, as opposed to the drier far-side slope, known as the _____

18. Moisture designations (two) for air masses include:_____

19. The 1995 Atlantic hurricane season featured_____ tropical storms, including _____ hurricanes. This places the year_____ in terms of overall records.

20. According to Table 8.2 and the Saffir–Simpson scale, what were the names of the category 4 and 5 hurricanes listed for 1995?_____

21. As air sinks toward the surface in a high-pressure system, its temperature and its relative humidity increase.

    a. true
    b. false

22. Urbanization, agriculture, pollution, and water diversion pose a greater ongoing threat to the Everglades than hurricanes do.

    a. true
    b. false

23. The lowest sea-level pressure ever recorded was

    a. off the coast of Florida
    b. in Death Valley, California
    c. in Siberia in winter
    d. in the central eyewall of Typhoon Tip

24. A well-developed, cP air mass in its source region would have which of the following characteristics?

    a. cold temperatures
    b. clear skies
    c. high pressure
    d. all of the above
    e. none of the above

25. The condensation process requires

    a. dew-point temperatures alone
    b. condensation nuclei and saturated air
    c. moisture droplets
    d. condensation nuclei alone

9

WATER RESOURCES

Water is not always naturally available when and where it is needed. From the maintenance of a house plant to the distribution of local water supplies, from an irrigation program on a farm to the rearrangement of river flows—all involve aspects of water budget and water-resource management. The availability of water is controlled by atmospheric processes and climate patterns.

This chapter begins by examining the hydrologic cycle, which is a model of the flows of water in the atmosphere, along the surface, and in the upper layers of Earth's crust. The water balance can be viewed from a global perspective (Figures 9.1 and 9.2) or for a specific area (Figure 9.3). The soil-water budget is an accounting of the hydrologic cycle for a specific area, with emphasis on plants and soil moisture. Figure 9.4 explains the elements of the water-balance equation.

The nature of groundwater is discussed and several examples are given of this abused resource. Groundwater resources are closely tied to surface-water budgets. We also consider the water withdrawn and consumed from available resources including groundwater and surface streams, in terms of both quantity and quality. Many aspects of this chapter may prove useful to you since we all interact with the hydrologic cycle on a daily basis. Surface water supplies, predominantly streams, represent an important water resource and are discussed in this chapter.

A water crisis looms ahead in many countries, especially in some regions—the Middle East, Africa, the Colorado River Basin, and Mexico, among others. Some 80 countries face impending water shortages, either in quantity or quality, or both. One billion people lack access to safe water in 2001; some 1.8 billion lack adequate sanitary facilities. During the first half of the new century water availability per person will drop by 74%, as population increases and adequate quality water decreases. A spatial understanding of

water resources is critical if we are to meet these challenges. The ☺ icon indicates that there is an accompanying animation on the Student CD. The ✺ icon indicates that there is an accompanying satellite or notebook animation on the Student CD.

OUTLINE HEADINGS AND KEY TERMS

The first-, second-, and third-order headings that divide Chapter 9 serve as an outline for your notes and studies. The key terms and concepts that appear **boldface** in the text are listed here under their appropriate heading in ***bold italics***. All these highlighted terms appear in the text glossary. Note the check-off box (❑) so you can mark your progress as you master each concept. These terms should be in your reading notes or used to prepare note cards.

The outline headings and terms for Chapter 9:

The Hydrologic Cycle

☺ Earth's Water and the Hydrologic Cycle

   ❑ *hydrologic cycle*

A Hydrologic Cycle Model

Surface Water

   ❑ *interception*
   ❑ *infiltration*
   ❑ *percolation*

Soil-Water Budget Concept

✺ Global Water Balance Components

   ❑ *soil-water budget*

The Soil-Water-Balance Equation

   Precipitation (PRECIP) Input

   ❑ *precipitation*
   ❑ *rain gauge*

GEWEX Continental-Scale International Project (GCIP):
http://www.ogp.noaa.gov/mpe/gapp/index.htm

Agricultural Research Service:
http://www.ars.usda.gov/

Drought:
http//:drought.unl.edu/

High Plains Aquifer:
http://ne.water.usgs.gov/highplains/hpactivities.html
http://webserver.cr.usgs.gov/nawqa/hpgw/
    HPGW_home.html

Snowy Mountains Hydro:
http://www.snowyhydro.com.au/

USGS water use studies:
http://water.usgs.gov/public/watuse/

Middle East Water Information Network:
http://www.columbia.edu/cu/lweb/indiv/mideast/cuvlm/
    water.html

KEY LEARNING CONCEPTS FOR CHAPTER 9

The following key learning concepts help guide your reading and comprehension efforts. The operative word is in *italics*. Use these carefully to guide your reading of the chapter and note that STEP 1 asks you to work with these concepts. These same learning concepts are used in organizing the summary and review section at the end of the chapter—grouping together definitions, a list of key terms, and review questions.

After reading the chapter and using this study guide, you should be able to:

- *Illustrate* the hydrologic cycle with a simple sketch and *label* it with definitions for each water pathway.
- *Relate* the importance of the water-budget concept to your understanding of the hydrologic cycle, water resources, and soil moisture for a specific location.
- *Construct* the water-balance equation as a way of accounting for the expenditures of water supply and *define* each of the components in the equation and their specific operation.
- *Describe* the nature of groundwater, and *define* the elements of the groundwater environment.
- *Identify* critical aspects of freshwater supplies for the future, and *cite* specific issues related to sectors of use, regions and countries, and potential remedies for any shortfalls.

✳ STEP 1: Critical Thinking Process

Using your interest and learning, and the following questions as guidelines <u>only</u>, briefly discuss your experience with this chapter. In examining your learning you need not go through each of these questions in detail, simply provide an overview of your critical thinking process as it relates to some aspect of this chapter.

- What did you know about the learning concept before you began?
- Which information sources did you use in your learning (text, class, other)?
- Were you able to complete the action stated in the learning concept? What did you learn?
- Are there any aspects of the concept about which you want to know more?

Critical Thinking and Chapter 9: _____

_____

_____

_____

_____

✳ STEP 2: Hydrologic Cycle

1. What is the purpose of budgeting water and, specifically, the water balance? What are the uses for this concept? Explain.

_____

_____

2. Using Figure 9.1 and the section in the text describing the hydrologic cycle, complete the information and labels on the following figure.

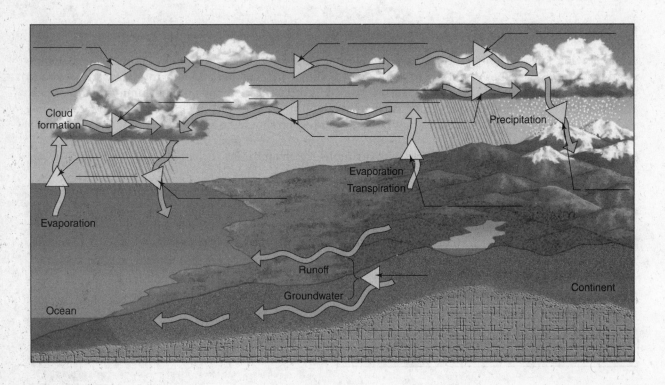

✳ STEP 3: Soil-Water Budget

1. Annual precipitation (PRECIP, water supply) for the United States and Canada is shown in Figure 9.6, annual potential evapotranspiration (POTET, water demand) is in Figure 9.8. Briefly compare these two maps and answer the following items.

(a) Can you identify from the two maps regions where PRECIP supply is _higher_ than POTET demand? Describe.

_____

_____

_____

(b) Can you identify from the two maps regions where POTET demand is *higher* than PRECIP supply? Describe.

_____

_____

_____

(c) Why do you think 95% of irrigated agriculture in the United States occurs west of the 95th meridian (central Kansas)?

_____

_____

_____

2. (a) Record the components of the water balance equation on the following line using the acronyms given in Figure 9.4.

_____ =(_____ – _____)+ _____ ± Δ _____

ACTET (Actual evapotranspiration)

(b) Define each of the following components of the water balance equation:

PRECIP: _____ ;

_____

POTET: _____

_____ ;

DEFIC: _____

_____ ;

SURPL: _____

_____ ;

±ΔSTRGE: _____

_____ ;

ACTET: _____

_____

Water balance data for Kingsport, Tennessee, to use in question #3.

Kingsport, Tennessee (Cfa): pop. 32,000, lat. 36° 30' N, long. 82° 30' W, elev. 391 m (1284 ft)

	Jan	Feb	Mar	Apr	May	Jun	Jul	Aug	Sep	Oct	Nov	Dec	Annual
Temperature°C	4.3	4.8	8.2	14.1	18.6	23.0	24.0	23.9	21.3	15.0	8.5	4.5	14.2
(°F)	(39.7)	(40.6)	(46.8)	(57.4)	(65.5)	(73.4)	(76.3)	(75.0)	(70.3)	(59.0)	(47.3)	(40.1)	(57.6)
PRECIP cm	9.7	9.9	9.7	8.4	10.4	9.7	13.2	11.2	6.6	6.6	6.6	9.9	111.9
(in.)	(3.8)	(3.9)	(3.8)	(3.3)	(4.1)	(3.8)	(5.2)	(4.4)	(2.6)	(2.6)	(2.6)	(3.9)	(44.1)
POTET cm	0.7	0.8	2.4	5.7	9.7	13.2	15.0	13.3	9.9	5.5	1.2	0.7	78.1
(in.)	(0.3)	(0.3)	(0.9)	(2.2)	(3.8)	(5.2)	(5.9)	(5.2)	(3.9)	(2.2)	(0.5)	(0.3)	(30.7)

3. Using the water-balance graph below, plot the Kingsport, Tennessee, data given above. Use a line graph for precipitation and a dashed line for potential evapotranspiration (you may use different colored pencils for these lines). Next, interpret the relationship between these demand and supply concepts by coloring in the spaces between the two plot lines: surplus, soil moisture utilization, deficit, and soil moisture recharge. Use the same color designations for each concept as in Figure 9.12.

When you finish the water balance graph, compare the months of March and September from Table 9.1 in the text with your graph. Check your graph and coloration of each component for those two months.

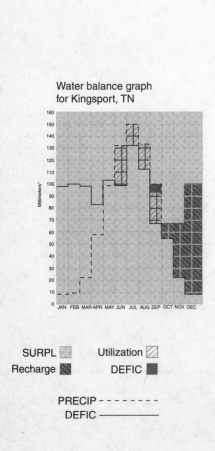

SURPL   Utilization
Recharge   DEFIC

PRECIP - - - - - - -
DEFIC ————

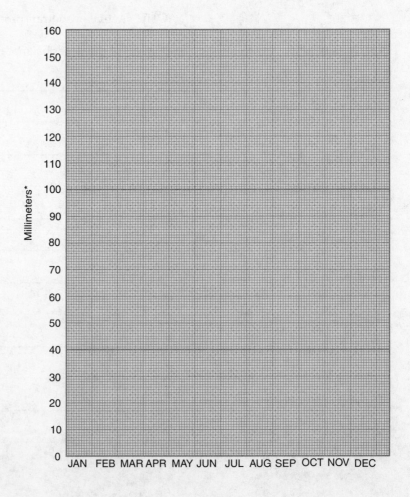

• 25.4 mm + 1 inch

4. Compare the water balance graph you have just completed with that for Phoenix, Arizona, presented in Figure 9.13e. What are the differences or similarities between Phoenix and Kingsport?

_____

_____

5. Compare Kingsport with Berkeley, California, in Figure 9.13a. What are the differences or similarities between Berkeley and Kingsport?

_____

_____

✳ STEP 4: Groundwater

1. The following illustration is derived from the right-hand half of Figure 9.17. Complete the labeling and descriptions to identify the various aspects of the groundwater environment. Use coloration to highlight the groundwater features in the illustration.

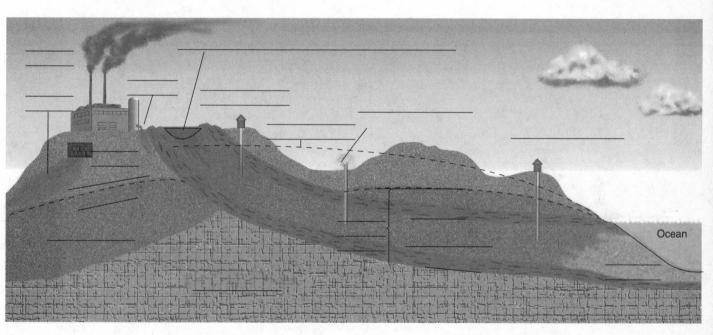

2. Analyze the evolving status of the High Plains Aquifer as presented in Focus Study 9.2. Then assume each of two points of view ("business as usual" and "strategies to halt the loss") and, given your analysis, respond to each point of view in considering the sustainability of the resource.

(a) Analysis of the situation: _____

_____

(b) "Business as usual"—no change in practices: _____

_____

_____

(c) "Strategies to halt groundwater mining"—change in practices: _____

_____

_____

_____

_____

✳ STEP 5: Our Water Supply

1. Using the following flow chart of the daily water budget for the lower 48 states of the United States, as shown in Figure 9.20, fill in the label for each cell and the quantity of water involved, in billions of gallons a day (BGD).

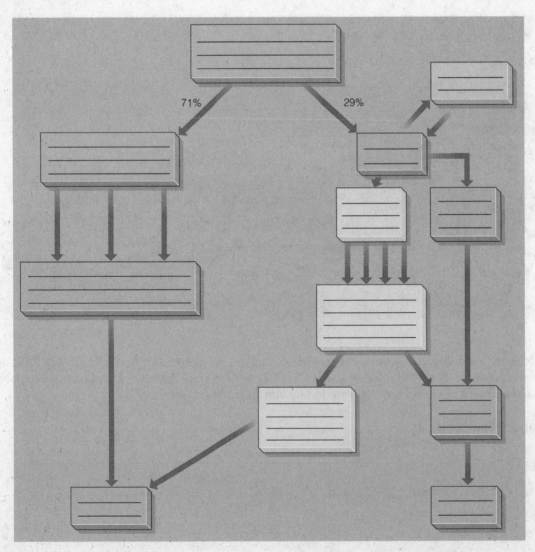

2. In examining the water withdrawal by sector pie charts on the map in Figure 9.21, compare water withdrawal sectors for the following areas. What is your general interpretation as to economic activity on each continent?

United States and Africa: _____

_____

_____

Canada and Asia: _____

_____

_____

3. What are some of the spatial and temporal obstacles to developing water resources?

_____

_____

_____

4. After looking at Table 9.2 how does the pattern of stream runoff compare with projected population growth? Which areas will continue to have surpluses? Which areas are facing shortages?

_____

_____

_____

_____

5. What are some of the coming water issues facing either China or the Middle East? What solutions have been proposed?

_____

_____

_____

_____

✳ STEP 6: NetWork—Internet Connection

There are many Internet addresses (URLs) listed in this chapter of the *Geosystems* textbook. Go to any two URLs, or "Destination" links on the *Geosystems* Home Page, and briefly describe what you find.

1. _____ : _____

_____

2. _____ : _____

_____
_____

SAMPLE SELF-TEST

(Answers appear at the end of the study guide.)

1. Most of the precipitation and evaporation on Earth takes place over the

    a. land masses
    b. oceans and seas
    c. poles of the planet
    d. subtropical latitudes

2. Relative to the hydrologic cycle, which of the following is <u>incorrect</u>?

    a. the bulk of the precipitation occurs over the ocean
    b. over 50 units of moisture are involved in advective flows in the hydrologic cycle model illustrated in the text
    c. 22 percent of Earth's precipitation falls over the land
    d. 78 percent of all precipitation falls on the oceans

3. Potential evapotranspiration refers to

    a. the moisture supply
    b. the amount of unmet water demand in an environment
    c. the amount of water that would evaporate or transpire if it were available
    d. the amount of water that only plants use

4. Precipitation is generally measured with

    a. buckets
    b. an evaporation pan
    c. computer technology
    d. a specially constructed rain gauge at over 100,000 places worldwide

5. ctual evapotranspiration is determined by

    a. PRECIP – DEFIC
    b. PRECIP – SURPL
    c. ACTET – DEFIC
    d. POTET – DEFIC

6. Water detention on the surface is a form of

    a. deficit
    b. groundwater
    c. surplus
    d. soil moisture storage

7. Soil moisture that plants are capable of accessing and utilizing is called

    a. wilting point water
    b. gravitational water
    c. available water
    d. hygroscopic water

8. The largest potential source of freshwater accessible in North America is

    a. groundwater
    b. ice sheets and glaciers
    c. stream discharge
    d. potential evapotranspiration

9. A water-bearing rock strata is called

    a. soil moisture storage
    b. an aquiclude
    c. a zone of aeration
    d. an aquifer

10. The upper limit of groundwater that is available for utilization at the surface is called

    a. capillary water
    b. an aquiclude
    c. the cone of depression
    d. the water table

11. Groundwater

    a. is seemingly unlimited when compared with surface supplies
    b. when polluted, is actually easier to clean up than is surface water
    c. should be considered separately from surface supplies
    d. is reduced by the mining of water

12. The Snowy Mountain project was designed to adjust water budgets over a region of

    a. the western United States
    b. Manitoba near the Nelson River
    c. the Great Dividing Range of New South Wales and Victoria, Australia
    d. the Canadian and American Rockies

13. Deficit (DEFIC) is the moisture demand in the water balance.

    a. true
    b. false

14. Precipitation specifically refers to rain, sleet, snow, and hail.

    a. true
    b. false

15. Moisture entering a soil body is referred to as soil moisture utilization.

    a. true
    b. false

16. The difference between field capacity and wilting point is called capillary water, almost all of which is available for extraction by plants and evaporation.

    a. true
    b. false

17. Permeability refers to the movement of water through soil or porous rock.

    a. true
    b. false

18. The rate of flow of a river is simply called its discharge.

    a. true
    b. false

10

GLOBAL CLIMATE SYSTEMS

Chapter 10 serves as a synthesis of content from Chapters 2 through 9—Parts One and Two of the text. Earth experiences an almost infinite variety of weather. Even the same location may go through periods of changing weather. This variability, when considered along with the average conditions at a place over time, constitutes climate. Climates are so diverse that no two places on Earth's surface experience exactly the same climatic conditions, although general similarities permit grouping and classification.

Scientists are studying and modeling global climate patterns to understand the dynamic changes that are occurring. Various physical indicators have led a consensus of scientists in the latest 2001 Intergovernmental Panel on Climate Change to agree that an anthropogenic warming is underway. This chapter concludes with a section that examines this exciting application of the science of physical geography to understanding climate change.

OUTLINE HEADINGS AND KEY TERMS

The first-, second-, and third-order headings that divide Chapter 10 serve as an outline for your notes and studies. The key terms and concepts that appear **boldface** in the text are listed here under their appropriate heading in ***bold italics***. All these highlighted terms appear in the text glossary. Note the check-off box (❏ so you can mark your progress as you master each concept. These terms should be in your reading notes or used to prepare note cards. The ☢ icon indicates that there is an accompanying animation on the Student CD. The ✿ icon indicates that there is an accompanying satellite or notebook animation on the Student CD.

The outline headings and terms for Chapter 10:

❏ *climate*

Earth's Climate System and Its Classification

✿ Global Climates: Genetic Map (Causal) Notebook
✿ Global Climates: Empirical Map (Köppen) Notebook
☢ Global Patterns of Precipitation

❏ *climatology*
❏ *climatic regions*

Climate Components: Insolation, Temperature, Pressure, Air Masses, and Precipitation

Classification of Climatic Regions

❏ *classification*
❏ *genetic classification*
❏ *empirical classification*

A Climate Classification System

Classification Categories

Global Climate Patterns

❏ *climograph*

Tropical Climates (equatorial and tropical latitudes)

Tropical Rain Forest Climates

Tropical Monsoon Climates

Tropical Savanna Climates

Mesothermal Climates (midlatitudes, mild winters)

Humid Subtropical Climates

Marine West Coast Climates

Mediterranean Dry-Summer Climates

Microthermal Climates (mid- and high-latitudes, cold winters)

Humid Continental Hot-Summer Climates

123

URLs listed in Chapter 10

Ice sources:
http://visibleearth.nasa.gov/cgi-bin/viewrecord?2172
http://www.natice.noaa.gov/

Kyoto Protocol:
http://www.unfccc.int/resource/kpstats.pdf

Intergovernmental Panel on Climate Change:
http://www.ipcc.ch/

U.S. Global Change Research Program:
http://www.usgcrp.gov/
http://globalchange.gov

Goddard Institute for Space Studies:
http://www.giss.nasa.gov/

Global Climate Observing System (GCOS):
http://www.wmo.ch/web/gcos/gcoshome.html

World Climate Research Program (WCRP):
http://www.wmo.ch/web/wcrp/wcrp-home.html

International Polar Year:
http://us-ipy.org

Global Hydrology and Climate Center:
http://www.ghcc.msfc.nasa.gov

Greenland Climate Network Project:
http://cires.colorado.edu/steffen/gc-net/gc-net.html

National Climate Data Center:
http://www.ncdc.noaa.gov/

National Environmental Satellite, Data, and Information Service:
http://www.nesdis.noaa.gov/

National Center for Atmospheric Research:
http://www.ncar.ucar.edu/

Pew Center on Global Climate Change:
http://www.pewclimate.org

For Canada, information and research is coordinated by Environment Canada:
http://www.ec.gc.ca/climate/
http://www.socc.uwaterloo.ca/

Climate Prediction Center:
http://www.ncep.noaa.gov/

El Niño/La Niña sources:
http://www.jpl.nasa.gov/elnino/
http://www.pmel.noaa.gov/toga-tao/el-nino/
  nino-home.html

United Nations and WMO:
http://www.unep.org
http://www.wmo.ch/

Mauna Loa Observatory:
http://cdiac.esd.ornl.gov/ftp/maunaloa-
  co2/maunaloa.co2

Other Links of Interest

Climate Change:
http://www.ucsusa.org
http://yosemite.epa.gov/oar/
globalwarming.nsf/content/
ClimateScienceFAQ.html

Sustainable Energy Paths:
http://www.ipsep.org/
http://www.rmi.org

KEY LEARNING CONCEPTS FOR CHAPTER 10

The following key learning concepts help guide your reading and comprehension efforts. The operative word is in *italics*. Use these carefully to guide your reading of the chapter and note that STEP 1 asks you to work with these concepts. These same learning concepts are used in organizing the summary and review section at the end of the chapter—grouping together definitions, a list of key terms, and review questions.
   After reading the chapter and using this study guide, you should be able to:

- *Define* climate and climatology, and *explain* the difference between climate and weather.
- *Review* the role of temperature, precipitation, air pressure, and air mass patterns used to establish climatic regions.
- *Review* the development of climate classification systems, and *compare* genetic and empirical systems as ways of classifying climate.
- *Describe* the principal climate classification categories other than deserts, and *locate* these regions on a world map.
- *Explain* the precipitation and moisture efficiency criteria used to determine the arid and semiarid climates, and *locate* them on a world map.
- *Outline* future climate patterns from forecasts presented, and *explain* the causes and potential consequences of climate change.

✳ STEP 1: Critical Thinking Process

Using your interest and learning, and the following questions as guidelines <u>only</u>, briefly discuss your experience with this chapter. In examining your learning you need not go through each of these questions in detail, simply provide an overview of your critical thinking process as it relates to some aspect of this chapter.

- What did you know about the learning concept before you began?
- Which information sources did you use in your learning (text, class, other)?
- Were you able to complete the action stated in the learning concept? What did you learn?
- Are there any aspects of the concept about which you want to know more?

Critical Thinking and Chapter 10: _____

_____

_____

_____

1. Distinguish between _weather_ and _climate_ by defining each.

(a) weather: _____

(b) climate: _____

✳ STEP 2: El Niño-Southern Oscillation (ENSO)

1. After reading Focus Study 10.1 describe the El Niño-Southern Oscillation (ENSO). Describe the main patterns involved. Are the effects of ENSO predictable and consistent? What does studying ENSO tell us about the complexity of Earth's climate system?

_____

_____

_____

_____

_____

2. How are ENSO and La Niña related? How are they similar and how are they different?

_____

_____

_____

3. What are typical conditions for ENSO and La Niña?

ENSO: _____

_____

La Niña: _____

_____

✳ STEP 3: Climate System Components

1. Using the general climate map in Figure 10.4 and the detailed climate map in Figure 10.5, determine the appropriate climate designation for your present location. If this is different from your hometown, also list the climate type for your home. See the "Geography I.D." page in the introduction to this study guide where you can also record this information.

Campus: _____

Hometown (if different): _____

2. Given your answer in question #1 (your climate classification) and using the following schematic of temperature and precipitation interactions from Figure 10.3, mark your present location (approximately) with an "**X**" on the figure below. Place the name of this location next to the mark. Next, mark the location (approximately) that is most characteristic of your birthplace with an "**O**" and record the name of this place next to the mark. (Consult with the world climate map in Figure 10.5.)

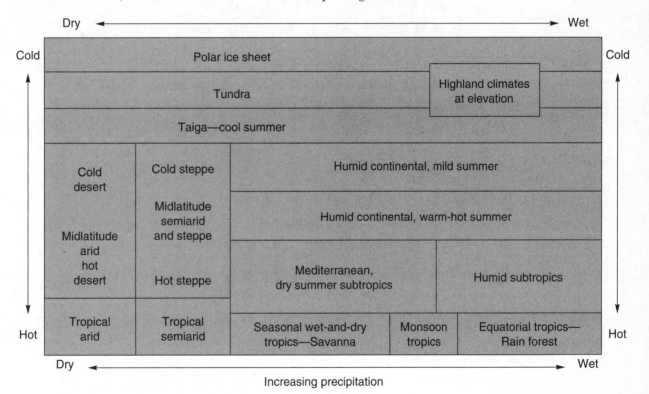

✳ STEP 4: Climate Classification

1. According to the text, what is the difference between an *empirical* and a *genetic classification*? Use climatic factors as an example.

_____

2. Use the four climographs on the next four pages and sets of data to plot: **mean monthly temperature** (*solid-line graph*), **mean monthly precipitation** (*a bar graph*), and **potential evapotranspiration** (*a dotted line*) data for the following stations. Use the Köppen climate guideline boxes in Appendix B to determine each station's classification. Analyze the distribution of temperature and precipitation during the year and record the other information as requested to complete each climograph. Analyze the distribution of temperature and precipitation during the year and, using your atlas and Figure B.1 in Appendix B, find and record the other information as requested to complete each climograph.

(a) Tropical climate: Salvador (Bahia), Brazil (#1)

(b) Mesothermal climate: New Orleans, Louisiana (#2)

(c) Microthermal climate: Montreal, Quebec, Canada (#3)

(d) Desert climate: Yuma, Arizona (#4)

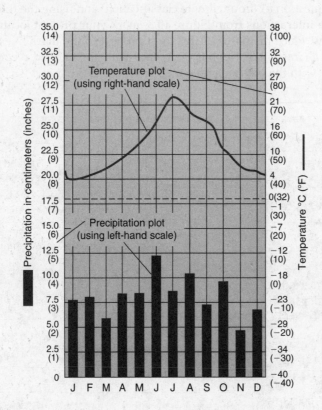

2a. Tropical climate: Salvador (Bahia), Brazil (#1)

Latitude _____

Longitude _____

Elevation _____

Population _____

Total annual rainfall: _____

Average annual temperature: _____

Annual temperature range: _____

Distribution of temperature during the year:

_____

_____

Distribution of precipitation during the year:

_____

_____

Principal atmospheric lifting mechanism(s):

_____

Distribution of potential evapotranspiration during

the year: _____

Salvador (Bahia), Brazil: pop. 1,507,000, lat. 12°59′S, long. 38°31′W, elev 9 m (30 ft)

	Jan	Feb	Mar	Apr	May	Jun	Jul	Aug	Sep	Oct	Nov	Dec	Annual
Temperature °C	26.0	26.3	26.3	25.8	24.8	23.8	23.0	22.9	23.6	24.5	25.1	25.6	24.8
(°F)	(78.8)	(79.3)	(79.3)	(78.4)	(76.6)	(74.8)	(73.4)	(73.2)	(74.5)	(76.1)	(77.2)	(78.1)	(76.6)
PRECIP cm	7.4	7.9	16.3	29.0	29.7	19.6	20.6	11.2	8.4	9.4	14.2	9.9	183.6
(in.)	(2.9)	(3.1)	(6.4)	(11.4)	(11.7)	(7.7)	(8.1)	(4.4)	(3.3)	(3.7)	(5.6)	(3.9)	(72.3)
POTET cm	13.7	12.4	13.5	11.9	10.7	9.1	8.6	8.6	9.4	11.2	11.9	13.2	134.4
(in.)	(5.4)	(4.9)	(5.3)	(4.7)	(4.2)	(3.6)	(3.4)	(3.4)	(3.7)	(4.4)	(4.7)	(5.2)	(52.9)

What are the main climatic influences for this station (air pressure, air mass sources, degree of continentality,

temperature of ocean currents)? _____

_____

_____

Köppen climate classification symbol: _____ ; name: _____

explanation for this determination:_____

_____

Representative biome (terrestrial ecosystem) characteristic of region: _____

Characteristic vegetation: _____

2b. Mersothermal climate: New Orleans, Louisiana (#2)

Latitude _____

Longitude _____

Elevation _____

Population _____

Total annual rainfall: _____

Average annual temperature: _____

Annual temperature range: _____

Distribution of temperature during the year:

_____

_____

Distribution of precipitation during the year:

_____

_____

Principal atmospheric lifting mechanism(s):

_____

_____

Distribution of potential evapotranspiration during the year: _____

_____

New Orleans, Louisiana: pop. 557,000, lat. 29°57'N, long. 90°04'W, elev. 3 m (9 ft)												
	Jan	Feb	Mar	Apr	May	Jun	Jul	Aug	Sep	Oct	Nov	Dec	Annual
Temperature °C	13.3	14.4	17.2	21.1	24.4	27.8	28.3	28.3	26.7	22.8	16.7	13.9	21.1
(°F)	(56.0)	(58.0)	(63.0)	(70.0)	(76.0)	(82.0)	(83.0)	(83.0)	(80.0)	(73.0)	(62.0)	(57.0)	(70.0)
PRECIP cm	12.2	10.7	16.8	13.7	13.7	14.2	18.0	16.3	14.7	9.4	10.2	11.7	161.3
(in.)	(4.8)	(4.2)	(6.6)	(5.4)	(5.4)	(5.6)	(7.1)	(6.4)	(5.8)	(3.7)	(4.0)	(4.6)	(63.5)
POTET cm	2.2	2.6	4.9	8.4	12.7	16.8	18.0	17.1	13.9	8.8	4.0	2.4	111.8
(in.)	(0.9)	(1.0)	(1.9)	(3.3)	(5.0)	(6.6)	(7.1)	(6.7)	(5.5)	(3.5)	(1.6)	(0.9)	(44.0)

What are the main climatic influences for this station (air pressure, air mass sources, degree of continentality, temperature of ocean currents)? _____

_____

_____

Köppen climate classification symbol: _____ ; name: _____

explanation for this determination: _____

_____

Representative biome (terrestrial ecosystem) characteristic of region: _____

Characteristic vegetation: _____

2c. **Microthermal** climate: Montreal, Quebec, Canada (#3)

Latitude _____

Longitude _____

Elevation _____

Population _____

Total annual rainfall: _____

Average annual temperature: _____

Annual temperature range: _____

Distribution of temperature during the year:

_____

_____

Distribution of precipitation during the year:

_____

_____

Principal atmospheric lifting mechanism(s):

_____

_____

Distribution of potential evapotranspiration during the year: _____

Montreal, Quebec, Canada: pop. 2,818,000, lat. 45°30′N, long. 73°35′W, elev. 57 m (187 ft)

	Jan	Feb	Mar	Apr	May	Jun	Jul	Aug	Sep	Oct	Nov	Dec	Annual
Temperature °C	−10.0	−9.4	−3.3	5.6	13.3	18.3	22.1	19.4	15.0	8.3	0.6	−6.7	6.0
(°F)	(14.0)	(15.1)	(26.1)	(42.1)	(55.9)	(64.9)	(70.0)	(66.9)	(59.0)	(46.9)	(33.1)	(19.9)	(42.8)
PRECIP cm	9.6	7.7	8.8	6.6	8.0	8.7	9.5	8.8	9.3	8.7	9.0	9.1	103.8
(in.)	(3.8)	(3.0)	(3.5)	(2.6)	(3.1)	(3.4)	(3.7)	(3.5)	(3.7)	(3.4)	(3.5)	(3.6)	(40.8)
POTET cm	0	0	0	2.7	8.1	11.9	13.9	12.1	7.7	3.7	0.2	0	60.3
(in.)	(0)	(0)	(0)	(1.1)	(3.2)	(4.7)	(5.5)	(4.8)	(3.0)	(1.5)	(0.1)	(0)	(23.7)

What are the main climatic influences for this station (air pressure, air mass sources, degree of continentality, temperature of ocean currents)? _____

_____

_____

Köppen climate classification symbol: _____; name: _____

 explanation for this determination: _____

_____

Representative biome (terrestrial ecosystem) characteristic of region: _____

Characteristic vegetation: _____

2d. Desert climate: Reno, Nevada (#4)

Latitude _____

Longitude _____

Elevation _____

Population _____

Total annual rainfall: _____

Average annual temperature: _____

Annual temperature range: _____

Distribution of temperature during the year:

_____

_____

Distribution of precipitation during the year:

_____

_____

Principal atmospheric lifting mechanism(s):

_____

_____

Distribution of potential evapotranspiration during
the year: _____

_____

Yuma, AZ: pop. 42,000, lat. 32°40'N, 114°36'W, elev. 61 m (200 ft)

	Jan	Feb	Mar	Apr	May	Jun	Jul	Aug	Sep	Oct	Nov	Dec	Annual
Temperature °C	1.7	14.8	17.9	21.5	25.1	29.6	33.2	32.8	29.7	23.3	17.1	13.2	22.6
(°F)	(54.9)	(58.6)	(64.2)	(70.7)	(77.2)	(85.3)	(91.8)	(91.0)	(85.5)	(73.9)	(62.8)	(55.8)	(72.7)
PRECIP cm	1.1	1.1	0.8	0.3	0.1	0	0.5	1.4	1.0	0.8	0.6	1.2	8.9
(in.)	(0.4)	(0.4)	(0.1)	(0.1)	(0)	(0)	(0.2)	(0.6)	(0.4)	(0.3)	(0.2)	(0.5)	(3.5)
POTET cm	1.3	2.3	4.6	8.2	13.6	18.9	21.1	20.0	16.4	9.0	3.4	1.5	110.3
(in.)	(0.5)	(0.9)	(1.8)	(3.2)	(5.4)	(7.4)	(8.3)	(7.9)	(6.5)	(3.5)	(1.3)	(0.6)	(43.4)

What are the main climatic influences for this station (air pressure, air mass sources, degree of continentality,
temperature of ocean currents)? _____

_____

_____

Köppen climate classification symbol:_____; name: _____

explanation for this determination: _____

_____

Representative biome (terrestrial ecosystem) characteristic of region: _____

Characteristic vegetation: _____

Note: *Geosystems* organizes the discussion of global climate patterns in special sections. *Tropical* climates begin with an introductory description of *Tropical* climates and a world map showing the distribution of these climates. A summary of the guidelines for this climate type appears at the end of the section for each major climate classification. *Mesothermal* climates begin in the next section following the tropical climates. The other climate classifications follow in the same manner.

3. What are the four types of arid and semiarid climates?

(a) _____

(b) _____

(c) _____

(d) _____

4. Briefly describe the polar environment. What are the three Polar climate classifications for these regions?

_____

_____

_____

5. Check the listings in the Index of the text for the following and briefly describe each from those sections in the text:

(a) Arctic region: _____

_____

(b) Antarctic region: _____

_____

(c) Periglacial landscapes: _____

_____

✳ STEP 5: Global Climate Change

1. What is a "GCM"? Use Figure 10.30 to assist you with your answer. Imagine a computer model that accurately portrays all the interconnections illustrated in Figure 10.1!

_____

_____

_____

2. What are the trends in the levels of carbon dioxide in the lower atmosphere (1774–2020 A.D. in Table 10.1)? What role does CO_2 play in the present warming trends?

_____

_____

_____

_____

3. What were the five warmest years in instrumental history (use the graph in Figure 10.28)? How do you characterize average temperatures during 1998–2003?

_____

_____

_____

4. Can you identify the cooling effect after the 1991 Mount Pinatubo eruption on atmospheric temperatures from the graph in Figure 10.29? If so, briefly describe it.

_____

_____

5. What countries and regions produce excessive CO_2 (Figure 10.29) for each of the following years in what percent (list in order from most to least):

(a) 1995: _____

(b) 2025 (forecast): _____

(c) What sectors in Figure 10.29 show the greatest increase in production over the 45-year period?

_____

_____

6. Identify and describe at least three potential consequences of a possible global warming.

(a) _____

_____

(b) _____

_____

(c) _____

_____

7. What are the high, middle, and low forecasts for this century, according to the 2001 Third Assessment Report from the IPCC? _____

_____

_____

8. After reading High Latitude Connection 10.1, list five changes that scientists have observed in the Arctic and Antarctic that indicate that we are experiencing global warming.

(a) _____

_____

(b) _____

_____

(c) _____

_____

(d) _____

_____

(e) _____

_____

9. What changes have indigenous peoples noticed in the Arctic? Why do you think they have been included in the ACIA process? _____

_____

_____

10. When the Intergovernmental Panel on Climate Change (IPCC) says that some policies to slow global change contain "no regrets" strategies (pp. 313–15), what do they mean?

_____

_____

_____

✳ STEP 6: NetWork—Internet Connection

There are many Internet addresses (URLs) listed in this chapter of the *Geosystems* textbook. Go to any two URLs, or "Destination" links on the *Geosystems* Home Page, and briefly describe what you find.

1. _____ : _____

_____

2. _____ : _____

_____

SAMPLE SELF-TEST
(Answers appear at the end of the study guide.)

1. An area that contains characteristic weather patterns is called a/an

    a. climatology
    b. El Niño, or ENSO
    c. weather phenomenon
    d. climatic region

2. An empirical classification is partially based on

    a. the interaction of air masses
    b. the origin or genesis of the climate
    c. mean annual temperature and precipitation
    d. causative factors

3. Relative to *tropical* climates, which of the following is <u>true</u>?

    a. all months average below 18°C (64.4°F)
    b. annual POTET exceeds PRECIP
    c. strong seasonality prevails
    d. all months average warmer than 18°C (64.4°F)

4. Relative to the *Mediterranean, hot summer* climates, which of the following is <u>false</u>?

    a. summers are hot
    b. 70% of the PRECIP occurs in the winter months
    c. they are found in California, South America, South Africa, the Mediterranean, and New Zealand
    d. its warmest summer month averages below 22°C (71.6°F)

5. The coldest climate on Earth, outside of the polar climates, is the

    a. *Tundra*
    b. *Humid continental, cold winter*

    c. *Ice Cap*

    d. *Subarctic*

6. Relative to an *Ice Cap* climate, which of the following is <u>correct</u>?

    a. the annual temperature range is less than 17 C° (30 F°)

    b. it is generally called the tundra

    c. its warmest month is below 0°C (32°F)

    d. its warmest month is above 0°C (32°F)

7. If PRECIP is more than 1/2 POTET but not equal to it, the climate is considered a

    a. *Tropical rain forest*

    b. *Desert*

    c. *Cold Desert*

    d. *Steppe*

8. The most extensive climates, occupying 36% of Earth's surface (land and ocean), are the

    a. Tropical climates

    b. Mesothermal climates

    c. Microthermal climates

    d. Polar climates

9. Which of the following is a typical *Subarctic, cool summer* climate?

    a. Churchill, Manitoba

    b. Lisbon, Portugal

    c. Dalian, China

    d. Verkhoyansk, Siberia, Russia

10. According to the 2001 IPCC assessment, temperatures will

    a. increase 1.4 C° to 5.8 C° (2.5 F° to 10.4 F°) in this century

    b. increase in a way unrelated to the behavior of society

    c. be 1.5 C° (2.7 F°) lower

    d. be 5.0 C° (9.0 F°) higher

11. Relative to future temperatures,

    a. humans cannot influence long-term temperature trends

    b. short-term changes appear to be out of our reach to influence

    c. a cooperative global network of weather monitoring among nations has yet to be established

    d. human society is causing short-term changes in global temperatures and temperature patterns

12. The warmest years in the history of weather instruments were

    a. recorded during the period 1910 to 1921

    b. between 1980 and 2003

    c. in the 1950s

    d. not determined since temperatures are not exhibiting any trend at this time

13. Climate is

    a. the weather of a region

    b. the short-term conditions of the atmosphere

    c. the long-term conditions in the atmosphere including extremes

    d. a reference to temperature patterns only

14. Hot and wet temperature and precipitation patterns are characteristic of

    a. rain forests in the equatorial tropics

    b. humid continental climates

    c. polar climates

    d. cold semiarid steppe climates

15. Which type of plants grow in desert climates?

    a. taiga

    b. boreal forests

    c. xerophytic

    d. chaparral

16. According to the quote (p. 309) from Richard Houghton and George Woodwell in *Scientific American*, climate zones are shifting and sea level is rising at this time.

    a. true

    b. false

17. The Intergovernmental Panel on Climate Change (IPCC) has reached unanimity among greenhouse experts that a climatic warming is occurring; uncertainty exists as to the severity and exact timing of consequences.

    a. true

    b. false

18. The Köppen climatic classification system is an example of a genetic classification.

    a. true

    b. false

19. The only Köppen classification that is based on moisture as well as temperature includes the H climates.

    a. true

    b. false

20. The Mesothermal climates make up the second-largest percentage of Earth's surface (land and water areas) and 55% of Earth's resident population.

    a. true

    b. false

21. Subarctic climates include regions of the highest degree of continentality on Earth.

    a. true

    b. false

22. POTET exceeds PRECIP in all parts of the B climates with no exceptions.

    a. true

    b. false

23. Most of the Sahara is characterized by the *low-latitude hot steppe* classification.

    a. true

    b. false

24. A climograph is a mechanical instrument used for measuring climates.

    a. true

    b. false

25. Using the climate map and the text, what is the climate where you live and attend college? _____

26. According to GCMs the polar regions are expected to experience a greater increase in temperatures than the tropics.

    a. true

    b. false

27. The United States is leading efforts to ratify the Kyoto protocol

    a. true

    b. false

PART THREE:
The Earth–Atmosphere Interface

OVERVIEW—PART THREE

Earth is a dynamic planet whose surface is actively shaped by physical agents of change. Part Three is organized around two broad systems of these agents—the endogenic, or internal, system, and the exogenic, or external, system. The *endogenic system* (Chapters 11 and 12) involves processes that produce flows of heat and material from deep below the crust and are powered by radioactive decay—this is the solid realm of Earth.

The *exogenic system* (Chapters 13–17) includes processes that set air, water, and ice into motion and are powered by solar energy—this is the fluid realm of Earth's environment. Thus, Earth's surface is the interface between two systems, one that builds the landscape and one that reduces it. Both are subjects of Part Three.

Name: _____   Class Section: _____

Date: _____   Score/Grade: _____

11

THE DYNAMIC PLANET

The twentieth century was a time of great discovery about Earth's internal structure and dynamic crust, yet much remains undiscovered. This is a time of revolution in our understanding of how the present arrangement of continents and oceans evolved. A new era of Earth-systems science is emerging, effectively combining various disciplines within the study of physical geography. The geographic essence of geology, geophysics, paleontology, seismology, and geomorphology are all integrated by geographers to produce an overall picture of Earth's surface environment.

OUTLINE HEADINGS AND KEY TERMS

The first-, second-, and third-order headings that divide Chapter 11 serve as an outline for your notes and studies. The key terms and concepts that

appear **boldface** in the text are listed here under their appropriate heading in ***bold italics***. All these highlighted terms appear in the text glossary. Note the check-off box (❑) so you can mark your progress as you master each concept. These terms should be in your reading notes or used to prepare note cards. The ✇ icon indicates that there is an accompanying animation on the Student CD. The ✹ icon indicates that there is an accompanying satellite or notebook animation on the Student CD.

The outline headings and terms for Chapter 11:

✹ **Earth's Varied Landscapes: Aerial Photo Gallery**

❑ *endogenic system*
❑ *exogenic system*
❑ *endogenic*

The Pace of Change

⚙ Applying Relative Dating Principles
- ❏ *geologic time scale*
- ❏ *uniformitarianism*

Earth's Structure and Internal Energy
- ❏ *seismic waves*

Earth's Core and Magnetism
- ❏ *[CT1]core*

Earth's Magnetism
- ❏ *geomagnetic reversal*

Earth's Mantle
- ❏ *mantle*
- ❏ *asthenosphere*

Earth's Lithosphere and Crust
- ❏ *crust*
- ❏ *Mohorovicic discontinuity (Moho)*
- ❏ *granite*
- ❏ *basalt*
- ❏ *isostasy*

The Geologic Cycle
❀ The Rock Cycle Notebook
❀ Table 11.2 Igneous Rocks Notebook
⚙ Formation of Intrusive Igneous Features
⚙ Foliation (Metamorphic Rock)
- ❏ *geologic cycle*

The Rock Cycle

Minerals and Rocks
- ❏ *mineral*
- ❏ *rock*
- ❏ *rock cycle*

Igneous Processes
- ❏ *igneous rock*
- ❏ *magma*
- ❏ *lava*

Intrusive and Extrusive Igneous Rocks
- ❏ *pluton*
- ❏ *batholith*

Classifying Igneous Rocks

Sedimentary Processes
- ❏ *sedimentary rocks*
- ❏ *lithification*
- ❏ *stratigraphy*

Clastic Sedimentary Rocks

Chemical Sedimentary Rocks
- ❏ *limestone*

Metamorphic Processes
- ❏ *metamorphic rock*

Plate Tectonics
⚙ Seafloor Spreading, Subduction
⚙ Pangaea Breakup, Plate Movements
❀ Plate Boundaries Notebook
⚙ India Collision with Asia
⚙ Transform Faults, Plate Margins

A Brief History
- ❏ *continental drift*
- ❏ *Pangaea*
- ❏ *plate tectonics*

Sea-Floor Spreading and Production of New Crust
- ❏ *sea-floor spreading*
- ❏ *mid-ocean ridges*

Subduction of the Lithosphere
- ❏ *subduction zone*

The Formation and Breakup of Pangaea

Plate Boundaries
- ❏ *transform faults*

Earthquake and Volcanic Activity

Hot Spots
- ❏ *hot spots*
- ❏ *geothermal energy*

SUMMARY AND REVIEW

News Reports, Focus Study, and High Latitude Connection

News Report 11.1: Radioactivity: Earth's Time Clock

News Report 11.2: Drilling the Crust to Record Depths

Focus Study 11.1: Heat from Earth—Geothermal Energy and Power

High Latitude Connection 11.1: Isostatic Rebound in Alaska

URLs listed in Chapter 11

Geologic time scale:

http://www.ucmp.berkeley.edu/exhibit/geology.html

Earth's interior:

http://www.solarviews.com/eng/earthint.htm

Rocks and minerals:
http://webmineral.com/
http://www.minsocam.org/MSA/Research_Links.html

Earth's interior and plate tectonics:
http://geology.usgs.gov/index.shtml/

Plate motion calculator:
http://www.ig.utexas.edu/research/projects/plates/plates.htm

Ocean drilling program:
http://www.icdp-online.de/
http://www-odp.tamu.edu/
http://www.oceandrilling.org/

Geothermal:
http://www.eren.doe.gov/geothermal/
http://geothermal.marin.org/
http://www.geothermal.org
http://www.smu.edu/geothermal/
http://geysers.com

KEY LEARNING CONCEPTS FOR CHAPTER 11

The following key learning concepts help guide your reading and comprehension efforts. The operative word is in *italics*. Use these carefully to guide your reading of the chapter and note that STEP 1 asks you to work with these concepts. These same learning concepts are used in organizing the summary and review section at the end of the chapter—grouping together definitions, a list of key terms, and review questions.

After reading the chapter and using this study guide, you should be able to:

- *Distinguish* between the endogenic and exogenic systems, *determine* the driving force for each, and *explain* the pace at which these systems operate.
- *Diagram* Earth's interior in cross section and *describe* each distinct layer.
- *Illustrate* the geologic cycle, and *relate* the rock cycle and rock types to endogenic and exogenic processes.
- *Describe* Pangaea and its breakup, and *relate* several physical proofs that crustal drifting is continuing today.
- *Portray* the pattern of Earth's major plates, and *relate* this pattern to the occurrence of earthquakes, volcanic activity, and hot spots.

✳ STEP 1: Critical Thinking Process

Using your interest and learning, and the following questions as guidelines <u>only</u>, briefly discuss your experience with this chapter. In examining your learning you need not go through each of these questions in detail, simply provide an overview of your critical thinking process as it relates to some aspect of this chapter.

- What did you know about the learning concept before you began?
- Which information sources did you use in your learning (text, class, other)?
- Were you able to complete the action stated in the learning concept? What did you learn?
- Are there any aspects of the concept about which you want to know more?

Critical Thinking and Chapter 11: _____

_____

_____

_____

_____

1. Using the outline of Figure 11.1 (below) of the *geologic time scale*, label each Eon, Era, Period, Epoch, and the date in millions of years before the present. Note several of the important events in Earth's life history noted in the figure.

Geologic Time Scale

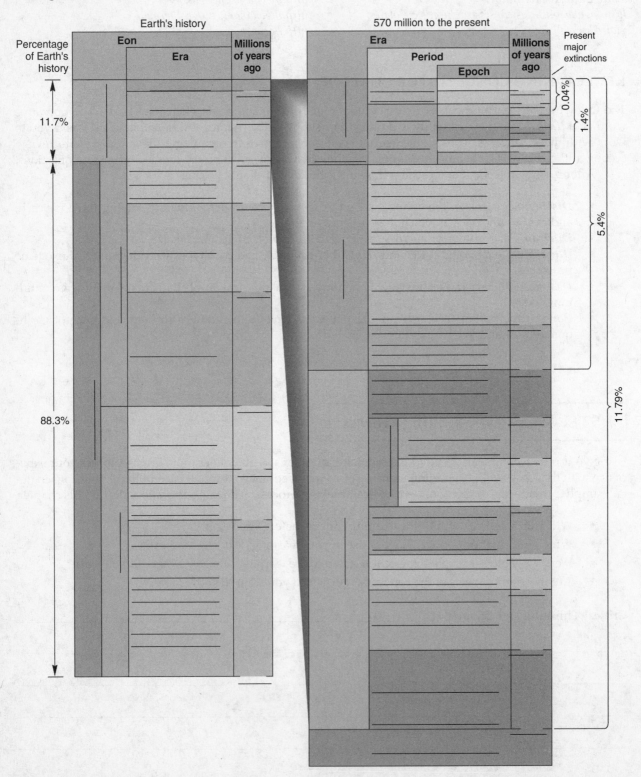

2. Place the following in the correct periods in the geologic time scale figure above:

(a) 1st land plants; insects
(b) 1st reptiles
(c) Beginning of age of dinosaurs
(d) Flowering plants
(e) 1st humans

3. List the names and dates of the six major extinction events. What makes the most recent event different from all the others? _____

_____

_____

_____

_____

4. According to the text, what two methods are used to determine the dates and ages of Earth's rocks?

(a) _____

(b) _____

❋ STEP 3: Earth's Structure

1. Using the following sketch derived from Figure 11.2b, detail the various layers within Earth. For each layer include its name and description (spaces on the left side) and depth below the surface in kilometers (spaces on the right side). The small wedge noted is detailed in the text in Figure 11.2c. You may want to use coloration with colored pencils. Place your labels in the spaces provided. (Note: Earth's profile is placed on a map of the United States and Canada in Figure 11.3 to give you a sense of scale and distance.)

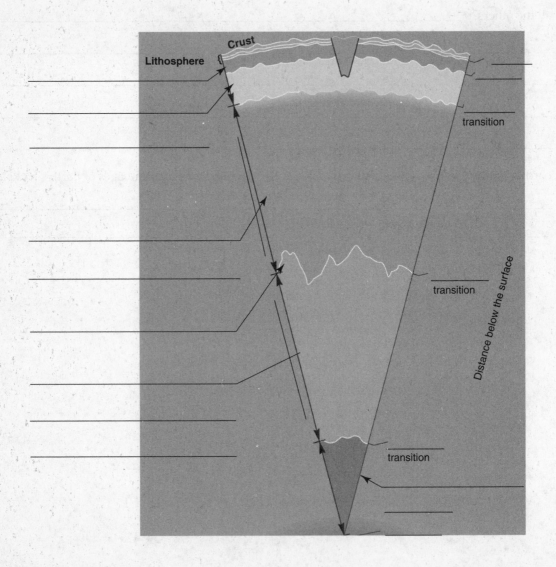

2. In your own words define and give examples of the following:

(a) endogenic system: _____

_____

(b) exogenic system: _____

_____

3. Relative to Earth's magnetic field, where is at least 90% of it generated? What relationship does it have to Earth's interior? What peculiar changes occur in the field that help scientists understand the crust? Explain.

_____

_____

_____

4. Explain what is meant by isostatic adjustment of the crust.

_____

_____

_____

5. After reading High Latitude Connection 11.1, discuss how global warming and the loss of glacial ice is related to the idea that the crust is in a constant state of compensating adjustment (see Figure 11.4).

_____

_____

_____

6. According to the text and illustrations (Figure 11.2, Figure 11.4, Figure 11.5), use the space below to sketch a cross section of continental and oceanic crust, Moho, uppermost mantle, and asthenosphere. Use coloration to highlight your sketch. Properly label each component.

7. What is the status of the various attempts to drill into the crust? Has anyone reached the Moho—the crust–mantle boundary—in drilling projects? (See News Report 11.2.)

_____

_____

✳ STEP 4: Geologic Cycle

1. Complete the labels on the following illustration identifying the various igneous forms and structures discussed in the text and Figure 11.7.

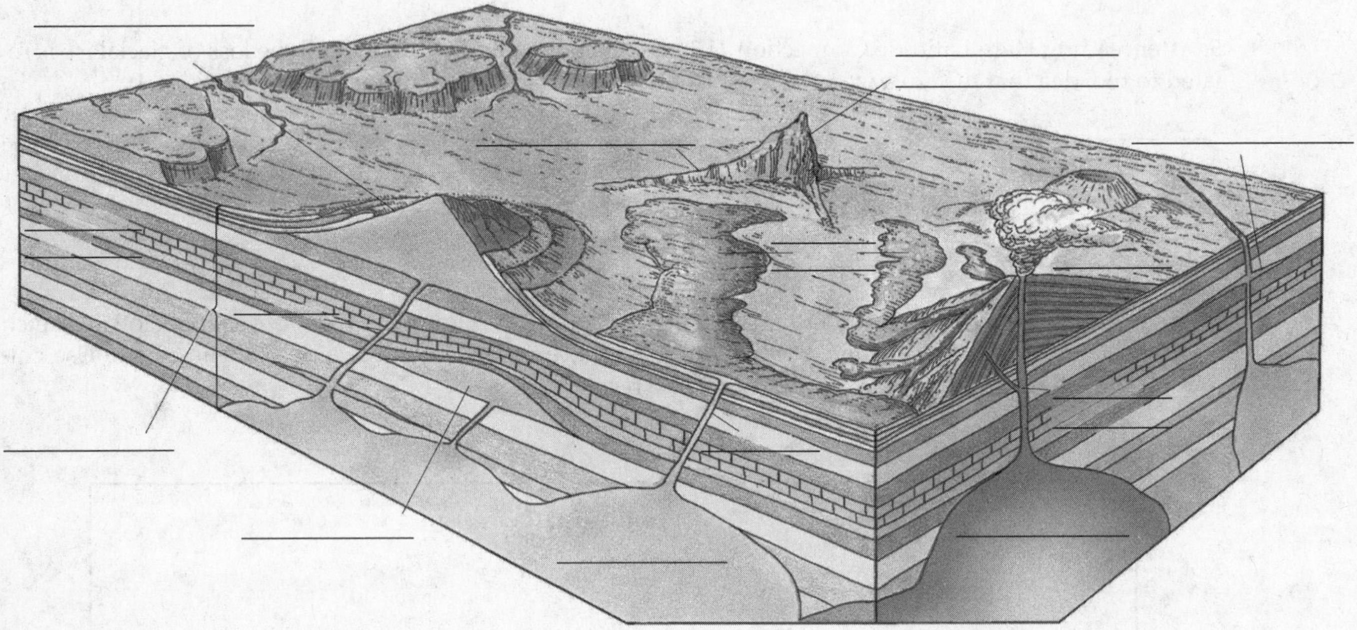

2. List the eight natural elements most common in Earth's crust; include the percentage of each (Table 11.1).

_____

_____

3. Briefly characterize the three rock forming processes described in the chapter.

(a) _____ : _____

_____

(b) _____ : _____

_____

(c) _____ : _____

_____

✳ STEP 5: Plate Tectonics

1. What does "Pangaea" mean? Who coined the name? Describe the processes that broke it apart. Are these crustal movements in operation today? _____

_____

_____

_____

_____

2. Using Figure 11.14, respond to the following:

(a) What is occurring beneath the Andes Mountains? _____

_____

(b) Describe the processes in operation in the center of the Atlantic Ocean basin. What does the remote sensing image disclose about this region? _____

_____

_____

3. In Figure 11.15, what is portrayed as occurring south of Iceland in the rocks and sediments of the ocean floor?

_____

_____

4. Label the 14 major lithospheric plates illustrated in Figure 11.18. Add additional arrows along with the ones presented indicating the direction of plate movement. Place your labels and directional arrows on the map below. If you lightly shade each plate with a different colored pencil it will make them easier to distinguish. Last: using red, shade those particular regions where earthquakes occur (see Figure 11.21).

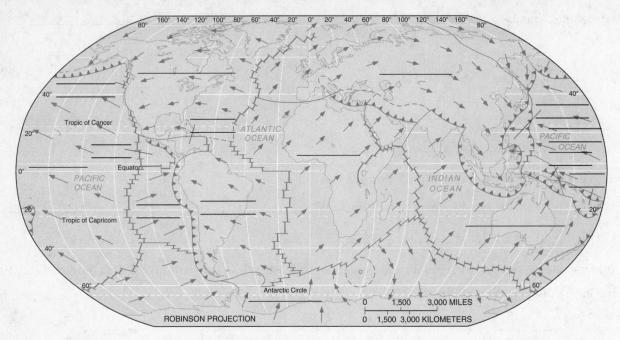

5. Examine the dramatic gravity anomaly map of the ocean floor in Figure 11.19. Can you find the features depicted in the image on the map in Figure 11.18 that appears just above the image? Now, take a moment and compare both of these to the chapter-opening map of the ocean floor for Chapter 12.

6. From the island of Hawaii to the island of Kauai, describe the ages of the islands from youngest to oldest (Figure 11.22). What process explains this age distribution?

_____

_____

_____

The age of Midway Island is:_____

❋ STEP 6: NetWork—Internet Connection

There are many Internet addresses (URLs) listed in this chapter of the *Geosystems* textbook. Go to any two URLs, or "Destination" links on the *Geosystems* Home Page, and briefly describe what you find.

1._____ : _____

_____

2._____ : _____

_____

SAMPLE SELF-TEST

(Answers appear at the end of the study guide.)

1. Which of the following is <u>endogenic</u> in nature?

    a. weathering

    b. erosion

    c. volcanism

    d. glaciers

    e. deposition

2. Of the following, which pair of concepts or terms are matched <u>correctly</u>?

    a. asthenosphere—lower mantle

    b. extrusive igneous—granitic crust

    c. upper mantle—Earth's magnetic field

    d. sea-floor spreading—mid-oceanic ridges

    e. subduction zone—upwelling and sea-floor spreading

3. Earth's crust is roughly made up of

    a. mantle and core material

    b. at least 14 major plates capable of movement

    c. strong unbroken material

    d. a thick layer at least 300 kilometers deep

    e. a brittle material that does not move

4. According to your text, the layer within Earth that lies directly <u>below</u> the lithosphere, but above the upper mantle, is best described as

    a. resistant to movement of any type

    b. a granitic material that weighs an average of 2.7 grams per cm^3

    c. a liquid nickel-iron composition

    d. a plastic-like layer that shatters if struck but flows when subjected to heat and pressure—known as the asthenosphere

5. Continental drift is an earlier term that describes

    a. motions that occurred about two billion years ago

    b. crustal plate movements, proposed by a geographer in 1912

    c. a brittle crust incapable of movement

    d. an old theory that has been disproved

6. Which of the following supports the plate tectonics concept?

    a. magnetic field patterns preserved in the rocks

    b. plant and animal fossil records

    c. radioactive decay dating of rocks on either side of a spreading center

    d. all of the above are true

    e. none of these is valid, since the theory has fallen into disfavor and has been dropped

7. The deepest single group of features of Earth's crust (continental or oceanic) are

    a. the areas of plate subduction beneath the ocean

    b. the mid-ocean ridge systems

    c. the deep lakes in East Africa

    d. the abyssal plains

8. Igneous rocks are the product of
   a. the accumulation of pieces of preexisting rocks
   b. heat and pressure causing hardening and physical changes in the rock
   c. weathering and erosion processes
   d. solidifying and crystallizing magma

9. The basic premise of catastrophism is that "the present is the key to the past."
   a. true
   b. false

10. The cementation, compaction, and hardening of sediments is called lithification.
    a. true
    b. false

11. The interior of Earth is known to science through <u>direct</u> physical observation and measurement in deep drill holes.
    a. true
    b. false

12. Earth's magnetic field is generated in the mantle and remains quite constant over time.
    a. true
    b. false

13. Seventy-five percent of Earth's crust is composed of only two elements—oxygen and silicon.
    a. true
    b. false

14. Rocks that solidify from a previous molten state are called metamorphic rocks.
    a. true
    b. false

15. Quartz (SiO_2) is generally higher in its resistance to weathering than are mafic minerals such as basalt.
    a. true
    b. false

16. Plate tectonics is regarded as the all-inclusive modern term for sea-floor spreading and subduction processes.
    a. true
    b. false

17. _____ assumes that *the same physical processes active in the environment today have been oper-ating throughout geologic time.* The phrase "the_____ is the key to the_____" is an expression coined to describe this principle. In contrast, the philosophy of_____ attempts to fit the vastness of Earth's age into a shortened time span.

18. A_____ is an element or combination of elements that forms an inorganic natural compound. Of the nearly 3000 minerals, only 20 are common, with just 8 of those making up_____% of the minerals in the crust (see Table 11.1). A_____ is an assemblage of minerals bound together or an aggregate of pieces of a single mineral.

19. Tectonic forces

    a. erode the surface of Earth
    b. warp, fold, and uplift rock
    c. control soil formation processes
    d. form sedimentary rock

20. The density of material below the Moho is_____ that above it.

    a. greater than
    b. less than
    c. the same as

21. What percentage of Earth's crust is composed of only eight natural elements?

    a. 25%
    b. 50%
    c. 70%
    d. 99%

22. Salt is an example of_____ rock

    a. sedimentary
    b. intrusive igneous
    c. extrusive igneous
    d. metamorphic

23. The Hawaiian Islands were formed as a result of

    a. a rising plume of magma from the mantle
    b. an oceanic–oceanic plate collision
    c. a continental–oceanic plate collision
    d. activity along a midocean ridge

24. Earthquake and volcanic occurrences do not correlate well with crustal plate boundaries.

    a. true
    b. false

12

TECTONICS, EARTHQUAKES, AND VOLCANISM

Tectonic activity has repeatedly deformed, recycled, and reshaped Earth's crust during its 4.6 billion year existence. The principal tectonic and volcanic zones lie along plate boundaries or in areas where Earth's crustal plates are affected by processes in the asthenosphere. The arrangement of continents and oceans, the origin of mountain ranges, and the locations of earthquake and volcanic activity are all the result of these dynamic endogenic processes.

Please use the chapter-opening illustration of "Floor of the Oceans" that opens Chapter 12 to summarize the principles of plate tectonics discussed in the last chapter. This dramatic portrait is meant as a bridge between Chapter 11 and this chapter. Follow the tour of the ocean floor printed below "Floor of the Oceans" and see if you can identify the features and places described.

The 1995 Kobe, Japan, earthquake caused $100 billion in damage, heartache, and a shock to the structure of Japanese society. Understanding these increasingly expensive hazards is an important goal of physical geography and its ability as a discipline to integrate human and an application of environmental themes.

OUTLINE HEADINGS AND KEY TERMS

The first-, second-, and third-order headings that divide Chapter 12 serve as an outline for your notes and studies. The key terms and concepts that appear **boldface** in the text are listed here under their appropriate heading in ***bold italics***. All these highlighted terms appear in the text glossary. Note the check-off box (❏) so you can mark your progress as you master each concept. These terms should be in your reading notes or used to prepare note cards. The ✸ icon indicates that there is an accompanying animation on the Student CD. The ❈ icon indicates

that there is an accompanying satellite or notebook animation on the Student CD.

The outline headings and terms for Chapter 12:

Earth's Surface Relief Features

❏ *relief*
❏ *topography*

Crustal Orders of Relief

First Order of Relief

❏ *continental platforms*
❏ *ocean basins*

Second Order of Relief

Third Order of Relief

Hypsometry

Earth's Topographic Regions

Crustal Formation Processes

✸ Terrane Formation

Continental Shields

❏ *continental shields*

Building Continental Crust and Terranes

❏ *terranes*

Crustal Deformation Processes

Folding and Broad Warping

✸ Folds, Anticlines, and Synclines
✸ Fault Types, Transform Faults, Plate Margins

❏ *folding*
❏ *anticline*
❏ *syncline*

Faulting

- ❏ *faulting*
- ❏ *earthquake*

Normal Fault

- ❏ *normal fault*

Reverse (Thrust) Fault

- ❏ *reverse fault*
- ❏ *thrust fault*

Strike-Slip Fault

- ❏ *strike-slip fault*

Faults in Concert

- ❏ *horst*
- ❏ *graben*

❈ Orogenesis (Mountain Building)

❈ Plate Boundaries Notebook

- ❏ *orogenesis*

Types of Orogenies

- ❏ *circum-Pacific belt*
- ❏ *ring of fire*

The Grand Tetons and the Sierra Nevada

The Appalachian Mountains

World Structural Regions

Earthquakes

✍ Seismograph, How It Works

✍ P- and S-Waves, Seismology

Expected Quakes and Those of Deadly Surprise

Focus, Epicenter, Foreshock, Aftershock

Earthquake Intensity and Magnitude

- ❏ *seismograph*

Moment Magnitude Scale Revises the Richter Scale

- ❏ *Richter scale*
- ❏ *moment magnitude scale*

The Nature of Faulting

- ❏ *elastic-rebound theory*

Earthquakes and the San Andreas Fault

Los Angeles Region

Earthquake Forecasting and Planning

Volcanism

✍ Volcanic Settings and Volcanic Activity

✍ Formation of Crater Lake

Volcanic Features

- ❏ *volcano*
- ❏ *crater*
- ❏ *lava*
- ❏ *pyroclastics*
- ❏ *aa*
- ❏ *pahoehoe*
- ❏ *cinder cone*
- ❏ *caldera*

Locations and Types of Volcanic Activity

Effusive Eruptions

- ❏ *effusive eruption*
- ❏ *shield volcano*
- ❏ *plateau basalts*

Explosive Eruptions

- ❏ *explosive eruption*
- ❏ *composite volcano*

Mount Pinatubo Eruption

Volcano Forecasting and Planning

SUMMARY AND REVIEW

News Reports, Focus Study, and Career Link

News Report 12.1: Mount Everest at New Heights

News Report 12.2: A Tragedy in Kobe, Japan—The Hyogo-ken Nanbu Earthquake

News Report 12.3: Seismic Gaps, Nervous Animals, Dilitancy, and Radon Gas

News Report 12.4: Is the Long Valley Caldera Next?

Focus Study 12.1: The 1980 Eruption of Mount St. Helens

Career Link 12.1: Travis Heggie, Geographer/Park Ranger/Ph.D.

URLs listed in Chapter 12

USGS Geology and Topography Map:
http://tapestry.usgs.gov

National Earthquake Information Center in Golden, Colorado:
http://wwwneic.cr.usgs.gov/

NRC, Earthquakes Canada:
http://www.seismo.nrcan.gc.ca/major_eq/ majoreq_e.php

Geohazards, earthquakes:
http://www.eqnet.org/
http://eqhazmaps.usgs.gov/
http://wwwneic.cr.usgs.gov/neis/bulletin/bulletin.html
http://www.colorado.edu/hazards/o/
http://www.colorado.edu/hazards/index.html
http://www.utoronto.ca/env/nh/toc.htm

National Geophysical Data Center:
http://www.ngdc.noaa.gov/seg/

Southern California Seismic Network:
http://www.trinet.org/scsn/scsn.html
http://www-socal.wr.usgs.gov/

Volcanoes:
http://volcano.und.nodak.edu/vwdocs/current_volcs/ current.html
http://vulcan.wr.usgs.gov/Volcanoes/framework.html
http://www.volcano.si.edu/gvp/
http://volcano.und.nodak.edu/
http://www.geo.mtu.edu/volcanoes/links/observatories. html

Long Valley Caldera:
http://quake.wr.usgs.gov/VOLCANOES/LongValley/

Mount St. Helens:
http://www.fs.fed.us/gpnf/mshnvm/volcanocam/
http://vulcan.wr.usgs.gov/
http://vulcan.wr.usgs.gov/Volcanoes/MSH/framework. html

Volcano Disaster Assistance Program:
http://vulcan.wr.usgs.gov/Vdap/framework.html

Volcano Cams from Around the World:
http://vulcan.wr.usgs.gov/Photo/volcano_cams.html

KEY LEARNING CONCEPTS FOR CHAPTER 12

The following key learning concepts help guide your reading and comprehension efforts. The operative word is in *italics*. Use these carefully to guide your reading of the chapter and note that STEP 1 asks you to work with these concepts. These same learning concepts are used in organizing the summary and review section at the end of the chapter—grouping together definitions, a list of key terms, and review questions.

After reading the chapter and using this study guide, you should be able to:

- *Describe* first, second, and third orders of relief, and *relate* examples of each from Earth's major topographic regions.
- *Describe* the several origins of continental crust, and *define* displaced terranes.
- *Explain* compressional processes and folding; *describe* four principal types of faults and their characteristic landforms.
- *Relate* the three types of plate collisions associated with orogenesis, and *identify* specific examples of each.
- *Explain* the nature of earthquakes, their measurement, and the nature of faulting.
- *Distinguish* between an effusive and an explosive volcanic eruption, and *describe* related landforms, using specific examples.

✳ STEP 1: Critical Thinking Process

Using your interest and learning, and the following questions as guidelines <u>only</u>, briefly discuss your experience with this chapter. In examining your learning you need not go through each of these questions in detail, simply provide an overview of your critical thinking process as it relates to some aspect of this chapter.

- What did you know about the learning concept before you began?
- Which information sources did you use in your learning (text, class, other)?
- Were you able to complete the action stated in the learning concept? What did you learn?
- Are there any aspects of the concept about which you want to know more?

_____

_____

_____

_____

❋ STEP 2: Surface Relief

1. Define each order of relief and give a <u>specific</u> example of each.

(a) first order: _____

(b) second order: _____

(c) third order: _____

2. Distinguish between the concepts of _relief_ and _topography_:

_____

_____

3. Describe in general terms the features of each type of topographic region mapped in Figure 12.3.

(a) plain: _____

_____

(b) high tableland: _____

_____

(c) hills and low tablelands: _____

_____

(d) mountains: _____

_____

(e) widely spaced mountains: _____

_____

(f) depressions: _____

_____

4. List Earth's nine major continental shields from Figure 12.4: _____

_____

_____

_____

✳ STEP 3: Crustal Formation and Deformation

1. Beginning at a sea-floor spreading center, describe the flow of magma to the surface, its subsequent movements, and the processes that eventually produce intrusive bodies of magma, continental crust, and sediment accumulations (see the illustration in Figure 12.5 and review from Chapter 11, Figure 11.5).

_____

_____

_____

2. What are terranes? What makes their composition different from the continent that surrounds them? Where is the Wrangellia Terrane (shown on the map in Figure 12.6a and discussed in the text) from and how far did it migrate from its source to its present location?

_____

_____

_____

_____

3. In the following illustration from Figure 12.7, identify each of the aspects of stress, strain, and surface expressions:

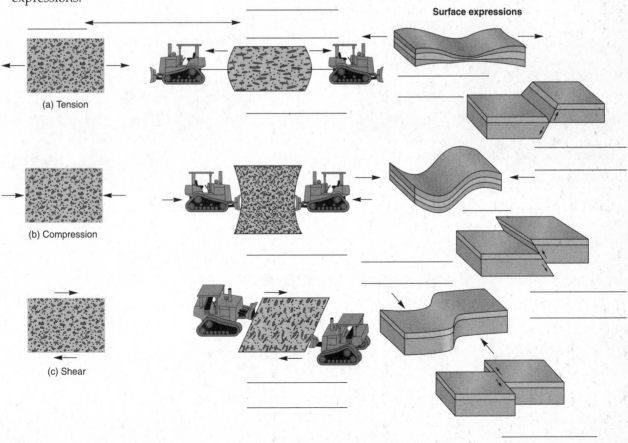

4. In the following illustration from Figure 12.8a, identify each of the components of this folded landscape. Circle the portion of the illustration that you think is depicted in the photograph in Figure 12.8b.

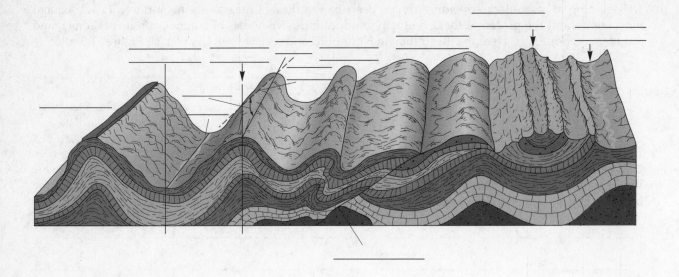

5. On the following illustration from Figure 12.11, identify each of the components of the basic types of faults and add the proper labels.

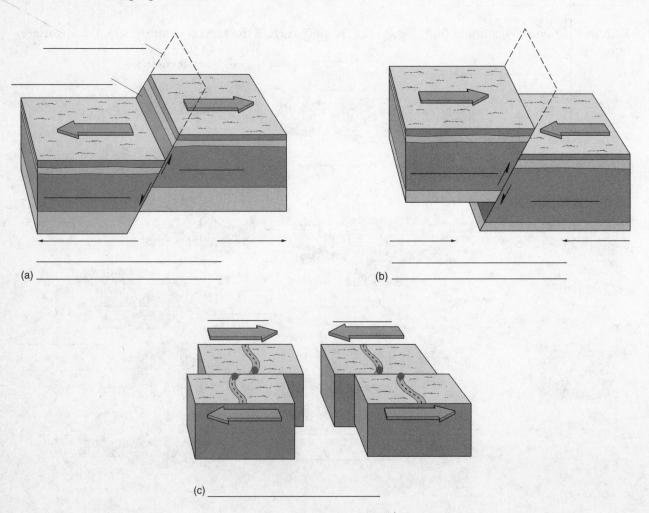

(a) _____

(b) _____

(c) _____

6. What are the three types of plate convergence mentioned in the text? What specific features do you associate with each type? Three insert maps of the ocean floor (taken from the chapter-opening map) are used as examples in Figure 12.16.

(a) _____

_____

(b) _____

_____

(c) _____

_____

✳ STEP 4: Earthquakes

1. Describe how earthquakes are measured and characterized in terms of *intensity* and *magnitude*.

(a) intensity: _____

_____

_____

_____

(b) magnitude (moment and amplitude): _____

_____

_____

2. In terms of earthquake characteristics and frequencies expected each year, complete the following from Table 12.1.

Characteristic Effects in Populated Areas	Approximate Intensity (modified Mercalli scale)	Approximate Magnitude (moment magnitude scale)	Number per year
Nearly total damage			
Great damage			
Considerable-to-serious			
Felt-by-all			
Felt-by-some			
Not felt, but recorded			

3. From the description in the text and Figure 12.20a, describe how the Loma Prieta earthquake was <u>not</u> characteristic of the San Andreas fault system even though it was produced along a portion of it.

_____

_____

_____

4. Relative to the San Andreas fault system define each of the following processes or motions as they relate to this system (Figure 12.12, the text, and review of Chapter 11, Figure 11.20).

(a) transform faults: _____

_____

(b) strike-slip: _____

_____

(c) right-lateral: _____

_____

5. The map of the seismicity of the United States and Canada in Figure 12.22 gives an indication of regions that most likely will experience earthquake activity in the future. Look at this map carefully. How do you assess your current location relative to the information on the map?

_____

_____

_____

_____

6. Briefly recount the impact of an earthquake prediction on a community as described by the chart in Figure 12.23. What interests might privately oppose such forecasting, according to the text?

_____

_____

_____

_____

✳ STEP 5: Volcanism

1. The text uses simplified criteria to classify volcanic eruptions—*effusive eruptions* and *explosive eruptions*. Describe the distinction between these two forms of volcanism, including the origin of magma, eruption characteristics, and landform features produced.

(a) effusive eruptions: _____

_____

_____

(b) explosive eruptions: _____

_____

_____

2. Refer back to the chapter-opening photo for Chapter 11. Which type of eruption is pictured?

_____

3. The principal mechanisms of volcanic activity are illustrated in Figure 12.25, which is reproduced here in part. Identify aspects of volcanic activity by placing the proper terms as labels in the spaces provided. Coloration may assist you in learning these components.

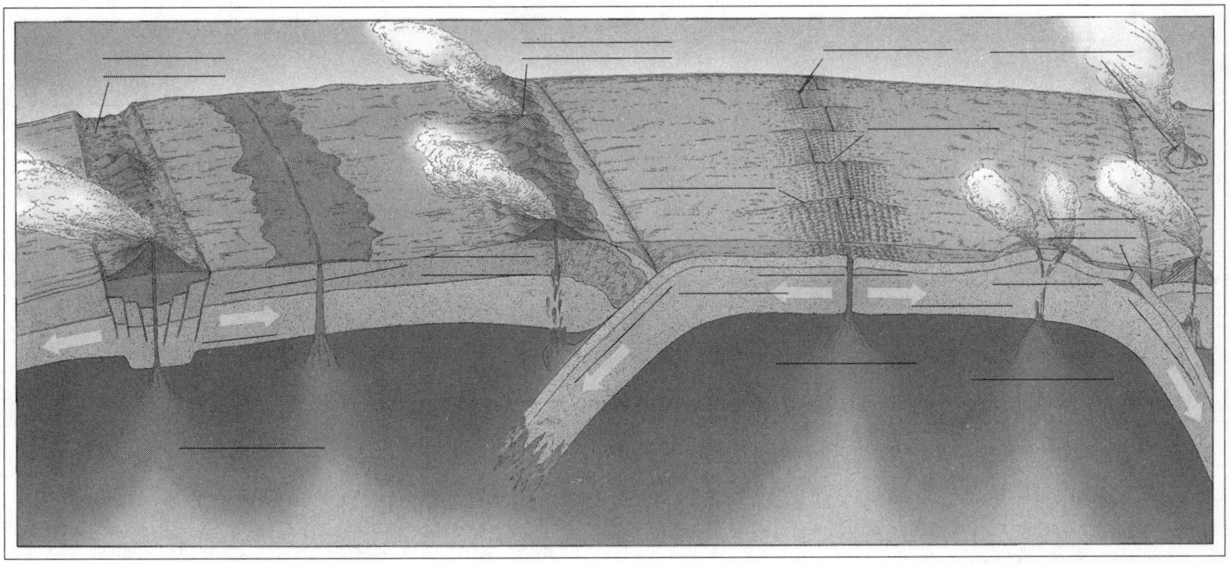

4. For the following locations, provide the type of eruption (explosive or effusive), the landform usually created (shield or composite volcano), the chemical and physical characteristics of the magma (mafic or felsic, fluid or thick and viscous), and the name of a volcano that is a good example.

(a) subduction boundary of a continental plate–oceanic plate convergence: _____

_____

_____

(b) sea-floor spreading center: _____

_____

_____

(c) hot spot: _____

_____

_____

5. Figure 12.34 shows the locations of extensive igneous provinces. What are these expanses? Give the names of the largest two provinces. Where are these largest ones? _____

_____

_____

✳ STEP 6: NetWork—Internet Connection

There are many Internet addresses (URLs) listed in this chapter of the *Geosystems* textbook. Go to any two URLs, or "Destination" links on the *Geosystems* Home Page, and briefly describe what you find.

1. _____ : _____

_____

2. _____ : _____

_____

SAMPLE SELF-TEST

(Answers appear at the end of the study guide.)

1. The deepest single group of features on the ocean floor are

  a. oceanic trenches
  b. mid-ocean ridge systems
  c. continental areas
  d. areas beneath the ice sheets
  e. orogenic belts

2. Which of the following represents a second order of relief?

  a. the North American plate
  b. the hills and valleys
  c. the Australian plate
  d. the Alps and the Rockies
  e. the oceanic basins

3. Terranes refer to

  a. the topography of a tract of land
  b. the "lay of the land"
  c. displaced and migrating pieces of Earth's crust
  d. features in the formation of Europe but not North America
  e. the continental shields within each continent

4. Faults that result from vertical displacements caused by crustal rifting of one portion of Earth's crust are called

   a. strike-slip, or transform faults
   b. reverse faults
   c. folded faults
   d. normal faults and reverse faults

5. Hawaiian volcanoes are

   a. explosive in nature
   b. related to subduction zones
   c. effusive and shield-like
   d. called composite
   e. presently dormant

6. The damage caused by the Loma Prieta earthquake

   a. was associated with a normal fault and not the San Andreas fault
   b. involved a portion of the San Andreas Fault in a strike-slip motion at depth
   c. was related to severe surface faulting and breaking
   d. produced ground displacements of as much as 21 feet
   e. was isolated to near the epicenter northeast of Santa Cruz

7. Mount St. Helens is a type of composite volcano associated with a subducted oceanic plate.

   a. true
   b. false

8. The most active tectonic regions of North and South America are on the

   a. western coasts
   b. eastern coasts
   c. southern coasts
   d. mid-continent
   e. areas near government offices

9. According to a chart in the text, the major impact of an earthquake prediction on a region would be

   a. a decline of and damage to the local economy
   b. increased business activity
   c. a slight but general increase in property values
   d. upward changes in employment opportunities
   e. massive evacuations due to an improvement in perception of the risk

10. Which of the following is matched <u>correctly</u>?

    a. anticline—a trough
    b. normal fault—horizontal motion
    c. reverse fault—overlying blocks that move upward relative to the footwall block
    d. syncline—a ridge
    e. folding—horsts and grabens

11. The ocean floor still remains somewhat of a mystery to modern science.

    a. true
    b. false

12. The Hawaiian Islands mark one end of a hot-spot track that stretches to Alaska.

    a. true
    b. false

13. Continental crust contains higher amounts of silica and aluminum and is composed of magma that is quite viscous (thick, sluggish in flow) when compared with magma directly from the asthenosphere.

   a. true
   b. false

14. Folding is principally associated with earthquake activity.

   a. true
   b. false

15. The Sierra Nevada and Grand Tetons are examples of folded mountain landscapes.

   a. true
   b. false

16. Earthquake damage may occur at distance from the epicenter, as was the case in the Mexico earthquake in 1985 and the Loma Prieta quake in 1989.

   a. true
   b. false

17. The variety of surface features on Earth results from

   a. tectonic activity
   b. gravity
   c. weathering and erosion
   d. all of these

18. The Grand Tetons best represents a

   a. horst
   b. graben
   c. tilted-fault block
   d. anticlinal structure

19. Molten rock that pours forth on Earth's surface is called

   a. metamorphic
   b. magma
   c. intrusive
   d. lava

20. Orogenesis can involve the capture of migrating terranes.

   a. true
   b. false

13

WEATHERING, KARST LANDSCAPES, AND MASS MOVEMENT

Chapter 13 begins the treatment of the external processes that affect Earth's surface. As the landscape is formed, a variety of processes simultaneously operate to wear it down, all under the influence of gravity.

The dynamics of slope and ever-changing adaptations and conditions of various slope elements produce a dynamic equilibrium among rock structure, climate, local relief, and elevation. Physical and chemical weathering operate toward the overall reduction of the landscape and the release of essential minerals from bedrock. Unique limestone landscapes occur in many regions of the world; these are the karst lands, caves, and massive caverns.

In addition, mass movement of surface material rearranges landforms, providing often-dramatic reminders of the power of nature. News accounts of mudflows, landslides, and the effects of weathering processes frequently hit the headlines. In July 1996, a 162,000-ton slab of granite fell over 600 m (2000 ft) into Yosemite Valley in Yosemite National Park—a sudden rockfall. The massive slab exploded into fine, powdery debris on impact, covering a large area in a light-gray blanket.

OUTLINE HEADINGS AND KEY TERMS

The first-, second-, and third-order headings that divide Chapter 13 serve as an outline for your notes and studies. The key terms and concepts that appear **boldface** in the text are listed here under their appropriate heading in *bold italics*. All these highlighted terms appear in the text glossary. Note the check-off box (❏) so you can mark your progress as you master each concept. These terms should be in your reading notes or used to prepare note cards.

The ✹ icon indicates that there is an accompanying satellite or notebook animation on the Student CD.

The outline headings and terms for Chapter 13:

Landmass Denudation

❏ *geomorphology*
❏ *denudation*
❏ *differential weathering*

Geomorphic Models of Landform Development

Dynamic Equilibrium View of Landforms

❏ *dynamic equilibrium model*
❏ *geomorphic threshold*

Slopes
❏ *slopes*

Weathering Processes

❏ *weathering*
❏ *regolith*
❏ *bedrock*
❏ *sediment*
❏ *parent material*

Factors Influencing Weathering Processes

❏ *joints*

Physical Weathering Processes

❏ *physical weathering*

Frost Action [CT2]

❏ *frost action*
❏ *talus slop*

Crystallization

Pressure-Release Jointing

- ❏ *sheeting*
- ❏ *exfoliation dome*

Chemical Weathering Processes

- ❏ *chemical weathering*
- ❏ *spheroidal weathering*

Hydration and Hydrolysis

- ❏ *hydration*
- ❏ *hydrolysis*

Oxidation

- ❏ *oxidation*

Carbonation and Solution

- ❏ *carbonation*

Karst Topography and Landscapes

❋ **Karst Farm Park, Indiana Notebook**

- ❏ *karst topography*

Formation of Karst

Lands Covered with Sinkholes

- ❏ *sinkholes*

Caves and Caverns

Mass Movement Processes

❋ **Madison River Landslide Notebook**

Mass Movement Mechanics

- ❏ *mass movement*
- ❏ *mass wasting*

The Role of Slopes

- ❏ *angle of repose*

Madison River Canyon Landslide

Classes of Mass Movements

Falls and Avalanches

- ❏ *rockfall*
- ❏ *debris avalanche*

Landslides

- ❏ *landslide*

Flows

- ❏ *mudflows*

Creep

- ❏ *soil creep*

Human-Induced Mass Movements
(Scarification)

- ❏ *scarification*

SUMMARY AND REVIEW

News Report, Focus Study, and Career Link

News Report 13.1: Amateurs Make Cave
Discoveries
Focus Study 13.1: Vaiont Reservoir Landslide
Disaster
Career Link: Gregory A. Pope, Geography
Professor

URLs listed in Chapter 13

**Natural Hazards Center at the University of
Colorado, Boulder:**
 http://www.Colorado.EDU/hazards/

Caves:
http://www.goodearthgraphics.com/virtcave/virtcave.
 html
http://hum.amu.edu.pl/?sgp/spec/links.html
http://fadr.msu.ru/~sigalov/ldlists.html

Natural hazards:
http://www.ngdc.noaa.gov/seg/hazard/hazards.html

Landslides:
http://landslides.usgs.gov/index.html
http://www.kingston.ac.uk/~ce_s011/slides.htm
http://anaheim-landslide.com/
http://landslides.usgs.gov/html_files/landslides/
 nationalmap/national.html

KEY LEARNING CONCEPTS FOR CHAPTER 13

The following key learning concepts help guide your reading and comprehension efforts. The operative word is in *italics*. Use these carefully to guide your reading of the chapter and note that STEP 1 asks you to work with these concepts. These same learning concepts are used in organizing the summary and review section at the end of the chapter—grouping together definitions, a list of key terms, and review questions.

After reading the chapter and using this study guide, you should be able to:

- *Define* the science of geomorphology.
- *Illustrate* the forces at work on materials residing on a slope.
- *Define* weathering, and *explain* the importance of the parent rock and joints and fractures in rock.
- *Describe* frost action, crystallization, pressure-release jointing, and the role of freezing water as physical weathering processes.
- *Describe* the susceptibility of different minerals to the chemical weathering processes called hydrolysis, oxidation, carbonation, and solution.
- *Review* the processes and features associated with karst topography.
- *Portray* the various types of mass movements, and *identify* examples of each in relation to moisture content and speed of movement.

✳ STEP 1: Critical Thinking Process

Using your interest and learning, and the following questions as guidelines <u>only</u>, briefly discuss your experience with this chapter. In examining your learning you need not go through each of these questions in detail, simply provide an overview of your critical thinking process as it relates to some aspect of this chapter.

- What did you know about the learning concept before you began?
- Which information sources did you use in your learning (text, class, other)?
- Were you able to complete the action stated in the learning concept? What did you learn?
- Are there any aspects of the concept about which you want to know more?

Critical Thinking and Chapter 13: _____

_____

_____

_____

_____

✳ STEP 2: Landmass Denudation Essentials

1. Define the following:

(a) geomorphology: _____

(b) denudation: _____

_____

(c) geomorphic threshold: _____

_____

2. Create an example that illustrates the four stages in the dynamic equilibrium model.

(a) equilibrium stability: _____

_____

(b) a destabilizing event: _____

_____

(c) period of adjustment: _____

_____

(d) development of a new and different condition of equilibrium stability: _____

_____

_____

3. Using the following line art from Figure 13.3a, complete the labels and components noted with leader lines that detail the forces acting on a slope and the principal elements of a typical slope form.

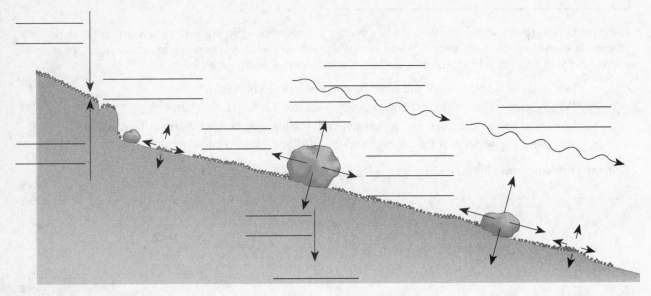

4. Using the following line art from Figure 13.3b, complete the labels and components noted with leader lines that detail the principal elements of a typical slope form.

Soil processes

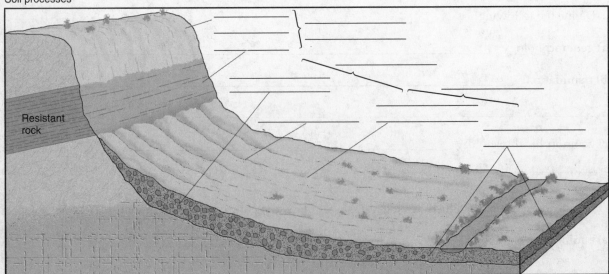

Resistant rock

✳ STEP 3: Weathering

1. The text describes the relationship among temperature, rainfall, and various types of weathering. The traditional model considered desert regions as places where chemical weathering ceased. What does the chapter state about weathering in dry regions at microsites on the rock? What is the importance of considering scale?

_____

_____

_____

2. Differentiate between *bedrock, regolith,* and *soil.* You may want to refer ahead to Chapter 18 when composing your soil response.

(a) bedrock: _____

_____

(b) regolith: _____

_____

(c) soil: _____

_____

3. Explain *pressure-release jointing* and the processes of *sheeting* and *exfoliation* as types of *physical weathering.* Can you identify these processes and features in the four photographs in Figure 13.11?

_____

_____

_____

4. Differentiate between *physical weathering* and *chemical weathering* processes. Give examples of each.

(a) Physical weathering: _____

_____ ;

example: _____

(b) Chemical weathering: _____

_____ ;

example: _____

5. Review the three photographs in Figure 13.12, and describe the processes that you identify operating in joints and on the rock formations. Review the caption and the text section in preparing your response.

_____

_____

_____

6. What process created the brilliant red colors in the rocks and soil shown in Figure 13.13? _____

_____

7. Answer the following relative to Figure 13.15 and Figure 13.16.

(a) What factors explain the pitted texture and indentations in the landscape? _____

_____

(b) What do you think happened to the disappearing streams noted in the illustration? _____

_____

(c) Describe the processes at work that produced the sinkhole pictured in Florida. _____

_____

8. How is the construction of the Arecibo radio telescope observatory (Figure 13.18) related to the characteristics of karst features? Explain. _____

_____

9. Relative to caves and caverns, define the following features and related processes that produce them (Figure 13.20). Examine the five photos that accompany the figure.

(a) stalactites: _____

_____

(b) stalagmites: _____

_____

(c) column: _____

_____

10. Identify with labels and descriptions the features of an underground cave as illustrated in Figure 13.20a.

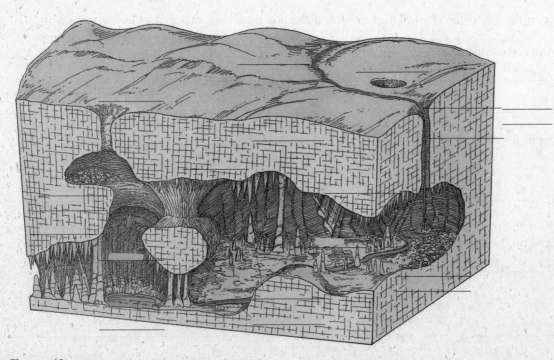

✳ STEP 4: Mass Movements

1. List the physical factors that contributed to the landslide event that occurred along the Madison River in 1959 (Figure 13.21). Include a mention of the triggering mechanism in your discussion.

_____

_____

2. Mass movement and mass wasting events are differentiated on the basis of water content and rates of movement. The following illustration is derived from Figure 13.21. Identify the type of movement depicted in each example and label them in the spaces provided.

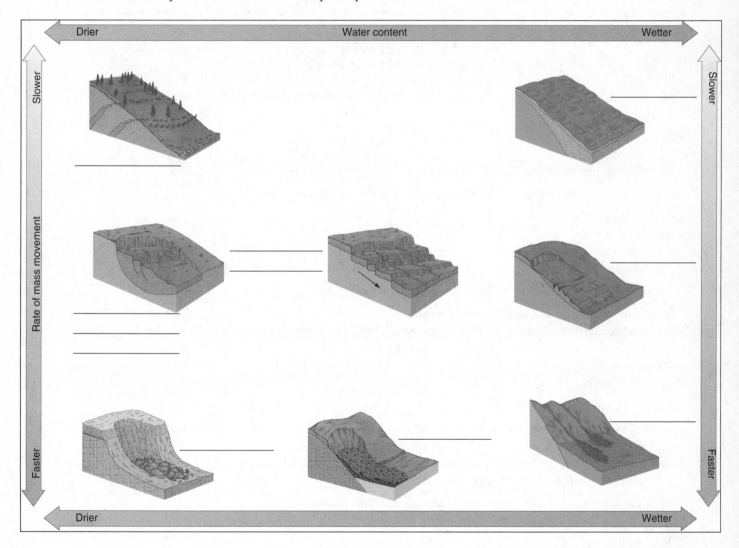

3. After reading Focus Study 13.1, the "Vaiont Reservoir Landslide Disaster," do you have any idea why the project went ahead as designed? What would you recommend to avoid such a disaster?

_____

_____

_____

There are many Internet addresses (URLs) listed in this chapter of the *Geosystems* textbook. Go to any two URLs, or "Destination" links on the *Geosystems* Home Page, and briefly describe what you find.

1. _____ : _____

    _____

2. _____ : _____

    _____

SAMPLE SELF-TEST

(Answers appear at the end of the study guide.)

1. The science that specifically studies the origin, evolution, form, and spatial distribution of landforms is

    a. geology
    b. geography
    c. geomorphology
    d. environmental chemistry

2. All processes that cause reduction and rearrangement of landforms are included in the term

    a. mass movement
    b. mass wasting
    c. weathering
    d. denudation

3. John Wesley Powell put forward the idea of base level, which refers to

    a. an evolutionary cycle of landscape development
    b. an imagined surface that extends inland from sea level, inclined gently upward
    c. a level below which a stream cannot erode its valley
    d. flat plateaus
    e. both b) and c) are correct

4. The dynamic equilibrium model refers to

    a. a balancing act between tectonic uplift and reduction rates of weathering and erosion in a given landscape
    b. a theory involving the cyclic or evolutionary development of a landscape
    c. landscapes that do not evidence ongoing adaptations to ever-changing conditions
    d. an important concept first stated by William Morris Davis

5. The disintegration and dissolving of surface and subsurface rock is called

    a. erosion
    b. mass wasting
    c. landmass denudation
    d. weathering

6. Chemical weathering is greatest under conditions of

    a. higher mean annual rainfall and temperatures
    b. lower mean annual rainfall and higher mean annual temperature
    c. lower mean annual rainfall and lower temperatures
    d. temperatures below freezing

7. An exfoliation dome in granitic rock forms through a process known as

   a. pressure-release jointing
   b. hydrolysis
   c. crystallization
   d. freeze-thaw action

8. Hydration is a process whereby a mineral

   a. chemically combines with water in chemical reactions
   b. dissolves in the presence of a weak acid
   c. swells upon the absorption of water, creating stress in rock
   d. oxidizes, most familiar in the "rusting" of iron

9. A local base level could be formed by a reservoir and the presence of a dam.

   a. true
   b. false

10. The geomorphic threshold is reached when a landform system possesses enough energy to overcome resistance against movement.

    a. true
    b. false

11. Slopes principally act as open material systems.

    a. true
    b. false

12. Regolith refers to partially weathered bedrock and may be missing or undeveloped in a region.

    a. true
    b. false

13. Freeze-thaw action of water in rocks is related to hydrogen bonding between water molecules and the expansion of water by as much as 9% in its volume as it cools and freezes.

    a. true
    b. false

14. Exfoliation domes and sheeting represent a form of pressure-release jointing.

    a. true
    b. false

15. Debris avalanches devastated the same villages in Peru twice within an eight-year period.

    a. true
    b. false

16. A lahar is a form of mass movement associated with snow avalanches.

    a. true
    b. false

17. The science that describes the origin, evolution, form, and spatial distribution of landforms is _____

18. Limestone is so abundant on Earth that many landscapes are composed of it. These landscapes comprise

    _____topography originally named for the _____ Plateau in

    Yugoslavia, where these processes were first studied.

19. A persistent mass movement of surface soil is called _____

20. Human-induced mass movements have created _____ of landscapes.

21. Human-induced mass movements produce a category of processes known as

    a. debris flow
    b. mass wasting
    c. scarification
    d. translational slides

22. The steepness of a slope of loose material at rest is the

    a. height of the slope
    b. angle of repose
    c. mass movement gradient
    d. the speed of individual particles

23. The text discusses spectacular tower karst topography as occurring in which country?

    a. Russia
    b. New Mexico, United States
    c. China
    d. France

24. The convex upper portion of a typical slope is called a

    a. pediment
    b. debris slope
    c. waxing slope
    d. free face

14

RIVER SYSTEMS AND LANDFORMS

This chapter begins with a discussion of the greatest rivers in the world in terms of flow discharge. Next, base level and the essential principles that govern stream flow are considered. The drainage basin is a basic hydrologic unit. A map presents these and the important continental divides for the United States and Canada. With this established, streamflow characteristics, gradient, and deposition are presented as water cascades through the hydrologic system.

The human component is irrevocably linked to streamflow as so many settlements are along river banks and on floodplains. Not only do rivers provide us with essential water supplies, but they also receive, dilute, and transport wastes and provide critical cooling water for industry. Rivers have been of fundamental importance throughout human history.

News Reports in this chapter highlight fascinating aspects of rivers and river management/mismanagement—dam releases to clear sediment out of the Grand Canyon, meandering rivers complicating boundaries, closing Niagara Falls for inspection, the Midwest floods, and a disappearing Nile delta.

This chapter discusses the dynamics of river systems and their landforms. Our successful interaction with river systems will be central to the sustainability of our economic behavior throughout the new century.

OUTLINE HEADINGS AND KEY TERMS

The first-, second-, and third-order headings that divide Chapter 14 serve as an outline for your notes and studies. The key terms and concepts that appear **boldface** in the text are listed here under their appropriate heading in ***bold italics***. All these highlighted terms appear in the text glossary. Note the check-off box (❑) so you can mark your progress as you master each concept. These terms should be in your reading notes or used to prepare note cards. The ✪ icon indicates that there is an accompanying animation on the Student CD. The ✿ icon indicates that there is an accompanying satellite or notebook animation on the Student CD.

The outline headings and terms for Chapter 14:

❑ *hydrology*

Fluvial Processes and Landscapes

  ❑ ***fluvial***

  ❑ ***erosion***

  ❑ ***transport***

  ❑ ***deposition***

  ❑ ***alluvium***

Base Level of Streams

  ❑ ***base level***

Drainage Basins

  ❑ ***drainage basin***

  ❑ ***watershed***

  ❑ ***sheetflow***

Drainage Divides and Basins

  ❑ ***continental divides***

Drainage Basins As Open Systems

Delaware River Basin

Internal Drainage

  ❑ ***internal drainage***

Drainage Density and Patterns

  ❑ ***drainage density***

  ❑ ***drainage pattern***

URLs listed in Chapter 14

Global Hydrology and Climate Center and Hydrology Web:
http://www.worldwater.org/links.htm
http://www.ghcc.msfc.nasa.gov/

The Amazon River:
http://boto.ocean.washington.edu/eos/index.html

The Danube River:
http://www2.unesco.org/mab/br/brdir/directory/
    biores.asp?code=ROM-UKR+01&mode=all
http://www.blacksea-environment.org/

Flood links:
http://www.dartmouth.edu/~floods/Resources.html
http://iwin.nws.noaa.gov/iwin/us/nationalflood.html

USGS river gauges:
http://water.usgs.gov/pubs/circ/circ1123/

Environment Canada's Water Survey of Canada:
http://www.msc-smc.ec.gc.ca/

Grand Canyon flooding experiment:
http://water.usgs.gov/pubs/FS/FS-060-99/
http://www.gcmrc.gov/

Midwest flood images:
http://rsd.gsfc.nasa.gov/rsd/images/Flood_cp.html
http://www.nwrfc.noaa.gov/floods/papers/oh_2/
    great.htm

http://www.floods.org/

http://www.dams.org/

Mississippi River:
http://www.mvn.usace.army.mil/pao/bcarre/
    bcarre.htm

Hydro-Québec project:
http://www.afn.ca/resolutions/1990/sca/res14.htm
http://www.carc.org/pubs/v20no2/1.htm

KEY LEARNING CONCEPTS FOR CHAPTER 14

The following key learning concepts help guide your reading and comprehension efforts. The operative word is in *italics*. Use these carefully to guide your reading of the chapter and note that STEP 1 asks you to work with these concepts. These same learning concepts are used in organizing the summary and review section at the end of the chapter—grouping together definitions, key terms, and review questions.

After reading the chapter and using this study guide, you should be able to:

- *Define* the term fluvial, and *outline* the fluvial processes: erosion, transportation, and deposition.
- *Construct* a basic drainage basin model, and *identify* different types of drainage patterns and internal drainage, with examples.
- *Describe* the relation among velocity, depth, width, and discharge, and *explain* the various ways that a stream erodes and transports its load.
- *Develop* a model of a meandering stream, including point bar, undercut bank, and cutoff, and *explain* the role of stream gradient in these flow characteristics.
- *Define* a floodplain, and *analyze* the behavior of a stream channel during a flood.
- *Differentiate* the several types of river deltas, and *detail* each.
- *Explain* flood probability estimates, and *review* strategies for mitigating flood hazards.

✴ STEP 1: Critical Thinking Process

Using your interest and learning, and the following questions as guidelines <u>only</u>, briefly discuss your experience with this chapter. In examining your learning you need not go through each of these questions in detail, simply provide an overview of your critical thinking process as it relates to some aspect of this chapter.

- What did you know about the learning concept before you began?
- Which information sources did you use in your learning (text, class, other)?
- Were you able to complete the action stated in the learning concept? What did you learn?
- Are there any aspects of the concept about which you want to know more?

Critical Thinking and Chapter 14:_____

_____

_____

_____

_____

✳ STEP 2: Fluvial Processes

1. Differentiate between the *ultimate base level* and a *local base level* (Figure 14.3a). Use examples to illustrate your descriptions.

(a) ultimate base level: _____

_____

(b) local base level: _____

_____

2. In terms of average discharge at their mouths, list the five greatest rivers on Earth, their discharge amounts, and the body of water into which they flow as listed in Table 14.1.

(a) _____

(b) _____

(c) _____

(d) _____

(e) _____

3. Complete the following from the text: "Water dislodges, dissolves, or removes surface material in a process called _____. Streams produce *fluvial erosion*, in which weathered sediment is picked up for _____ to new locations. A stream is a mixture of water and solids—carried in solution, suspension, and by mechanical _____. Materials are laid down by another process, _____ _____. _____ is the general term for the clay, silt, and sand deposited by running water."

4. List the five principal drainage orientations that are divided by continental divides in the United States and Canada (Figure 14.5).

(a) _____

(b) _____

(c) _____

(d) _____

(e) _____

5. Using the following map of the United States and Canada, label the major river *drainage basins* as identified in Figure 14.5. Also, add color lines to your map that designate the major *continental divides* that separate these drainage basins (consult the same map for reference).

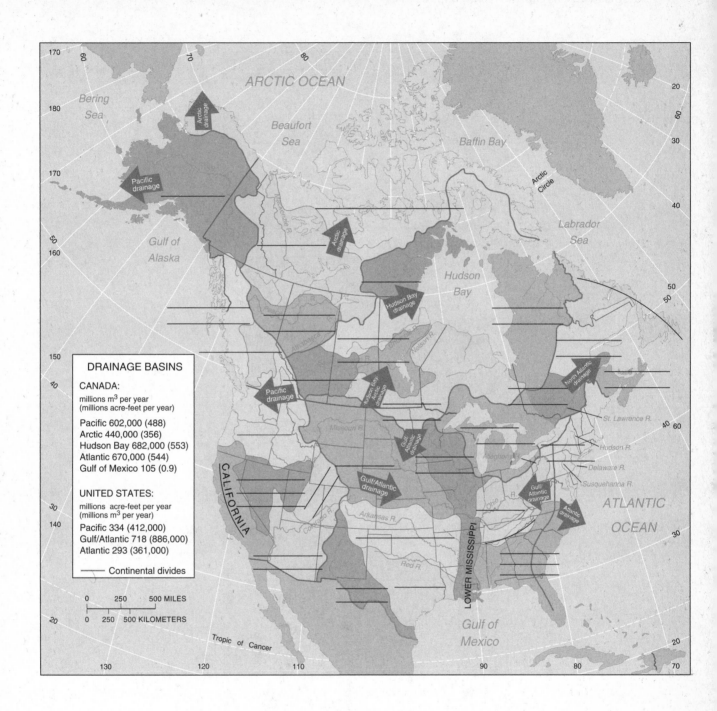

DRAINAGE BASINS

CANADA:
millions m³ per year
(millions acre-feet per year)

Pacific 602,000 (488)
Arctic 440,000 (356)
Hudson Bay 682,000 (553)
Atlantic 670,000 (544)
Gulf of Mexico 105 (0.9)

UNITED STATES:
millions acre-feet per year
(millions m³ per year)

Pacific 334 (412,000)
Gulf/Atlantic 718 (886,000)
Atlantic 293 (361,000)

—— Continental divides

0   250   500 MILES
0   250   500 KILOMETERS

6. Identify and label each of the following drainage patterns (from Figure 14.9). Then mark next to each illustration the appropriate figure number which is characteristic of: **(a)** the region shown in Figure 14.8a and b; and, **(b)** the Appalachian Mountain region shown by the image in Chapter 12, Figure 12.18a. **(c)** Mark with red the stream and its capture pictured in the inset.

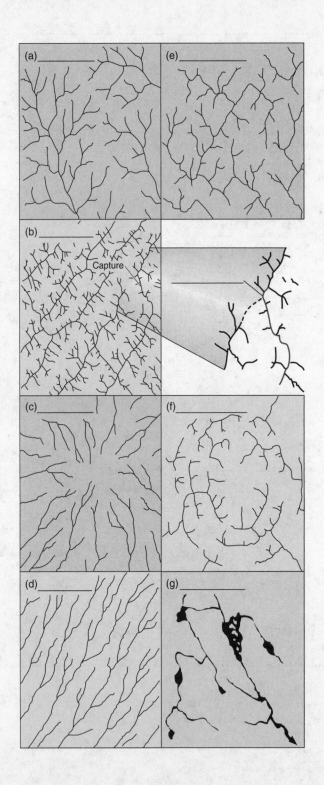

✳ STEP 3: Streamflow Characteristics

1. Differentiate between stream *competence* and *capacity* as they relate to stream transport ability.

(a) competence_____

_____

(b) capacity_____

_____

2. Briefly describe the ways in which a stream attacks its bed and conducts *erosion*.

_____

_____

_____

3. Briefly describe the ways in which a stream *transports* its load.

_____

_____

_____

4. Relative to the flood stages of the San Juan River channel in Utah depicted in Figure 14.10, what was occurring in terms of gauge height, discharge, and channel shape on each of the following dates:

(a) September 9, 1941: _____

_____

(b) October 14, 1941: _____

_____

(c) October 26, 1941: _____

_____

5. Analyze the actions and processes that are occurring in the two photos, Figure 14.12a and b. Based on the text discussion, include information about the energy and work going on in each scene.

(a) Figure 14.12a: _____

_____

_____

(b) Figure 14.12b: _____

_____

_____

6. How do the photo and satellite images in Figure 14.14 illustrate aspects of stream competence and capacity? Describe the load in the stream in Figure 14.14b. _____

_____

_____

7. In the space provided, diagram a stream cross section with an ideal *longitudinal profile* that illustrates a typical stream gradient. Add to the stream profile a *nickpoint* created by resistant rock strata that forms a waterfall (see Figures 14.18, 14.20, and 14.21).

8. Figure 14.16 depicts the development of river meanders and News Report 14.2, Figure 14.2.1, presents an example in which a former river meander is still in use as a political border between Nebraska and Iowa. Using the discussion in the text, answer the following. (The legend for topographic map symbols is Appendix A in the text).

(a) What is a river meander? Why do they form? _____

_____

_____

(b) Describe the development of an oxbow lake: _____

_____

_____

(c) Explain the sequence of events that produced the present circumstances surrounding Carter Lake, Iowa. Consult the map in News Report 14.2, Figure 14.2.1, and the discussion in the text as you prepare your answer.

_____

_____

_____

✳ STEP 4: Floods and River Management

1. On the following illustration of a typical floodplain, derived from Figure 14.23a, label the components in the appropriate spaces noted.

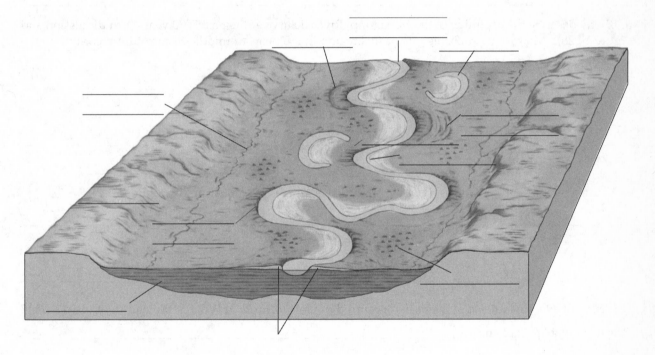

2. The topographic map in Figure 14.23d is of the Philipp, Mississippi, area. The area shown on the map segment is a portion of the lower Mississippi River floodplain. Using this map and the discussion in the text, answer the following. (The legend for topographic map symbols is Appendix A in the text.)

(a) The contour interval for this map segment is 5 feet.

(b) Are there levees along the river? If so, can you tell how high they are and how they are portrayed? ____

_____

(c) How many oxbow lakes do you count? _____

(d) Can you identify any meander scars on the landscape? Briefly describe them. _____

_____

_____

3. Briefly overview the details of the 1993 Midwest flood and related weather conditions (News Report 14.3, and text). What lessons were learned concerning floodplains, zoning, levee management, etc.?

_____

_____

_____

4. Describe the relationship between a depositional floodplain and the development of stream terraces. How might stream terraces (paired and unpaired) help in the interpretation of the stream's development in a valley (see text discussion and Figure 14.25)?

_____

_____

5. Briefly describe the evolution of the Mississippi River delta over the past 5000 years (text discussion and Figure 14.28). What is one probable fate of this region in the near- to middle-term future according to the text?

_____

_____

_____

_____

6. What is the present thinking as to why the Nile River delta is disappearing (News Report 14.4)?

_____

_____

_____

7. Figure 14.29b illustrates a problem that occurred in North Carolina. What is the nature of the disaster? Was it avoidable? Explain. _____

_____

_____

8. News Report 14.5 discusses the human history of Bayou Lafourche. What natural landscape features provided hurricane protection prior to human development and alteration of the bayou? What hurricane and flood protection measures have been taken to protect the bayou? What are the hurricane and flood issues facing Bayou Lafourche as a result of human activity? _____

_____

_____

_____

_____

9. What is meant by "Settlement Control Beats Flood Control" as discussed near the end of Focus Study 14.1?

_____

_____

_____

10. Label the following typical stream hydrograph for a drainage basin, taking care to identify those portions that apply to urbanization, both before and after development (Figure 14.29a).

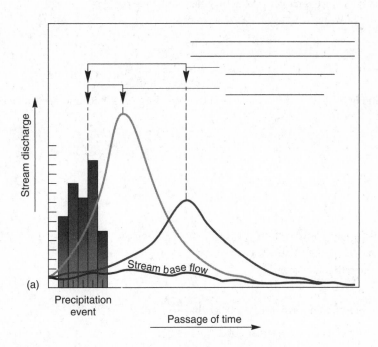

✳ STEP 5: NetWork—Internet Connection

There are many Internet addresses (URLs) listed in this chapter of the *Geosystems* textbook. Go to any two URLs, or "Destination" links on the *Geosystems* Home Page, and briefly describe what you find.

1. _____ : _____

_____

2. _____ : _____

_____

SAMPLE SELF-TEST

(Answers appear at the end of the study guide.)

1. Watersheds are defined by

    a. continental divides
    b. drainage divides
    c. unit hydrographs
    d. stream orders

2. Which of the following is <u>incorrectly</u> matched?

    a. rills—deep stream valleys
    b. gullies—developed rills
    c. drainage divides—ridges that control drainage
    d. Delaware River basin—in parts of five states

3. The Valley and Ridge Province is characterized by which drainage pattern?

    a. dendritic
    b. trellis
    c. parallel
    d. annular

4. The area depicted in Figure 14.7 near the junction of the West Virginia, Ohio, and Kentucky borders is characterized by which drainage pattern?

    a. dendritic
    b. trellis
    c. parallel
    d. annular

5. The Amazon River has

    a. the greatest discharge of any river on Earth
    b. not produced a delta in the Atlantic Ocean
    c. a braided mouth with many islands
    d. all of these are correct

6. The suspended load of a stream consists of particles that

    a. are rolled and bounced along the stream bed
    b. are held aloft in the stream flow
    c. drag along the stream bed
    d. are basically in solution

7. The downstream portion of a river

    a. generally becomes more sluggish
    b. generally is of higher velocity, which is masked by reduced turbulence
    c. usually has turbulent flows
    d. has lower discharges than do upstream portions

8. Relative to a meandering stream,

    a. the inner portion of a meander features a cut bank
    b. the inner portion of a meander features a point bar
    c. they tend to be straight
    d. they tend to develop when a steep gradient is formed

9. Niagara Falls is an example of a nickpoint related to differing resistances of bedrock strata.

   a. true
   b. false

10. Excess sediment in a stream will produce a maze of interconnected channels causing the stream to be braided.

    a. true
    b. false

11. Relative to stream gradient, it is not possible to have both ungraded and graded portions on the same stream.

    a. true
    b. false

12. The area of maximum velocity in a stream is usually in the inside of a bend near the point bar.

    a. true
    b. false

13. Natural levees are actually formed as by-products of flooding even though people depend on them for possible protection.

    a. true
    b. false

14. An oxbow lake is a form of braided stream.

    a. true
    b. false

15. The Nile River, which flows into the Mediterranean Sea, is an example of an arcuate delta.

    a. true
    b. false

16. Streamflows are measured with a staff gauge and stilling well, among other devices and techniques.

    a. true
    b. false

17. Urbanization both delays and lessens peak flow as plotted on a hydrograph.

    a. true
    b. false

18. The Mississippi-Missouri-Ohio river system is a single drainage basin rather than many smaller drainage basins, each considered separately.

    a. true
    b. false

19. Relative to continental divides, the Delaware River flows into the _____, the Mississippi River flows into the _____, the Columbia River flows into the ____, and the Yukon River flows into the _____

20. Stream drainage patterns associated with steep slopes and some relief are usually _____; those with a faulted and jointed landscape are _____; those with volcanic peaks are _____; and those associated with structural domes and concentric rock patterns are _____

21. The _____ is now blocked off from the Mississippi River at the Old Control Structure, yet is one-half the distance of the present Mississippi River channel to the Gulf of Mexico and is the site for a potential future channel change.

22. During the past 5000 years, the Mississippi River delta has assumed _____ (number of) deltaic forms. The present bird-foot delta has been building for the past _____ years at least.

23. Drainage divides are

    a. the same as continental divides
    b. ridges that divide watersheds from each other
    c. rills
    d. similar to nickpoints

24. The largest-sized material that can be carried by a stream is carried as

    a. bed load
    b. suspended load
    c. dissolved load
    d. truck loads

25. As a stream approaches its base level, the ability of the stream to erode its bed

    a. increases
    b. decreases
    c. stays about the same

26. If a stream has just enough gradient and discharge to transport its sediment load, the stream is said to be

    a. fitted
    b. competent
    c. in a nonequilibrium state
    d. graded

27. Natural levees are formed by

    a. floods
    b. normal flow conditions
    c. low flow conditions
    d. none of the above

15

EOLIAN PROCESSES AND ARID LANDSCAPES

Wind is an agent of erosion, transportation, and deposition. Its effectiveness has been the subject of much debate; in fact, wind at times was thought to produce major landforms. Presently, wind is regarded as a relatively minor exogenic agent, but it is significant enough to deserve our attention. In this chapter we examine the work of wind, its associated processes, and resulting landforms.

In addition, we discuss desert landscapes, where water remains the major erosional force but where an overall lack of moisture and stabilizing vegetation allows wind processes to operate. Evidence of this includes sand seas and sand dunes of infinite variety. Beach and coastal dunes, which form in many climates, also are influenced by wind and are discussed in this chapter.

OUTLINE HEADINGS AND KEY TERMS

The first-, second-, and third-order headings that divide Chapter 15 serve as an outline for your notes and studies. The key terms and concepts that appear **boldface** in the text are listed here under their appropriate heading in *bold italics*. All these highlighted terms appear in the text glossary. Note the check-off box (❏) so you can mark your progress as you master each concept. These terms should be in your reading notes or used to prepare note cards.

The outline headings and terms for Chapter 15:

The Work of Wind

❏ *eolian*

Eolian Erosion

❏ *deflation*
❏ *abrasion*

Deflation

❏ *desert pavement*
❏ *blowout depressions*

Abrasion

❏ *ventifacts*
❏ *yardangs*

Eolian Transportation

❏ *surface creep*

Eolian Depositional Landforms

❏ *dune*
❏ *erg desert*
❏ *sand sea*
❏ *slipface*

Loess Deposits

❏ *loess*

Overview of Desert Landscapes

Desert Climates
Desert Fluvial Processes
❏ *flash flood*
❏ *wash*
❏ *playa*
❏ *alluvial fan*
❏ *bajada*

Desert Landscapes

Basin and Range Province

❏ *Basin and Range Province*
❏ *bolson*

Desertification

❏ *desertification*

SUMMARY AND REVIEW

News Report and Focus Study

News Report 15.1: The Dust Bowl

Focus Study 15.1: The Colorado River: A System
Out of Balance

URLs listed in Chapter 15

Surviving the Dust Bowl:
http://www.pbs.org/wgbh/amex/dustbowl/

Desertification:
http://www.unccd.int/main.php.
http://www.undp.org/drylands/

KEY LEARNING CONCEPTS FOR CHAPTER 15

The following key learning concepts help guide your reading and comprehension efforts. The operative
word is in *italics*. Use these carefully to guide your reading of the chapter and note that STEP 1 asks you to
work with these concepts. These same learning concepts are used in organizing the summary and review
section at the end of the chapter—grouping together definitions, a list of key terms, and review questions.

After reading the chapter and using this study guide, you should be able to:

- *Characterize* the unique work accomplished by wind and eolian processes.
- *Describe* eolian erosion, including deflation, abrasion, and the resultant landforms.
- *Describe* eolian transportation, and *explain* saltation and surface creep.
- *Identify* the major classes of sand dunes, and *present* examples within each class.
- *Define* loess deposits and their origins, locations, and landforms.
- *Portray* desert landscapes, and *locate* these regions on a world map.

✳ STEP 1: Critical Thinking Process

Using your interest and learning, and the following questions as guidelines <u>only</u>, briefly discuss your experience with this chapter. In examining your learning you need not go through each of these questions in detail, simply provide an overview of your critical thinking process as it relates to some aspect of this chapter.

- What did you know about the learning concept before you began?
- Which information sources did you use in your learning (text, class, other)?
- Were you able to complete the action stated in the learning concept? What did you learn?
- Are there any aspects of the concept about which you want to know more?

Critical Thinking and Chapter 15: _____

_____

_____

_____

_____

_____

_____

✳ STEP 2: Work of Wind

1. Describe the relationship between sand movement and wind velocity as measured over a meter-wide strip of ground surface (see Figure 15.1 in the text). _____

_____

_____

_____

2. Briefly explain the relationship between the Persian Gulf War (1991), ORVs, and deflation rates in desert regions as mentioned in the text._____

_____

_____

_____

3. An extensive area of sand and sand dunes is called an _____ desert or a _____. In contrast most of Earth's deserts comprise desert pavement; some provincial (local) names for these pavements include (give at least three with locations):

    (a)_____

    (b)_____

    (c)_____

4. In terms of eolian erosion and transport, what is shown in the photograph in Figure 15.5, and discussed in the text, that relates to human impact and intervention with coastal dunes?

_____

_____

_____

5. Briefly compare and contrast *eolian transportation* in this chapter and *fluvial transportation* (discussed in Chapter 14).

_____

_____

_____

_____

6. Define *loess* and describe both the glacial and the desert sources and processes that produce these materials. Refer to the photographs in Figure 15.13a and b. _____

_____

_____

_____

✳ STEP 3: Deserts and Arid Lands

1. Label the cross-section illustration of a sand dune that appears in Figure 15.9.

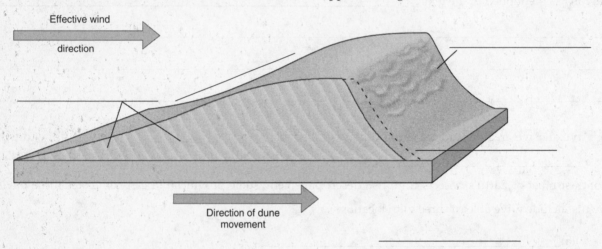

Effective wind direction

Direction of dune movement

2. Label each type of sand dune pictured below as derived from Figure 15.10, and draw an arrow showing the effective or prevailing wind direction involved in the formation of each.

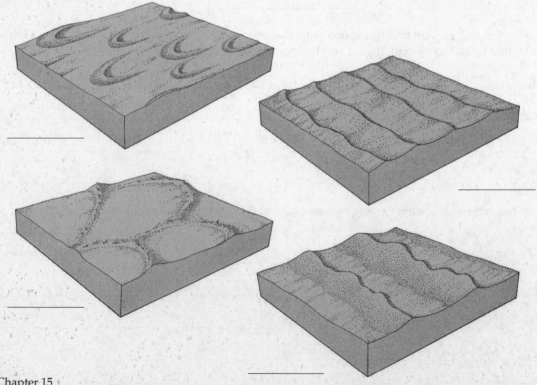

3. Although rare, a precipitation event in the desert is often dramatic. Describe what occurs following a downpour in the desert (examine Figure 15.17a and b). _____

_____

_____

4. Using the arid-lands map, Figure 15.15, and the text explanation, describe the processes that produce this spatial distribution of deserts and steppes—detail at least three locations on Earth and related desert-producing processes. Be sure and take a moment and compare the different photos that accompany the map. Earth's deserts possess a variety of appearances and forms.

(a) _____

_____;

(b) _____

_____;

(c) _____

_____

5. Figure 15.18 presents a photograph and a topographic map of an alluvial fan—a typical landform feature in a desert environment. (The legend for topographic map symbols is Appendix A in the text.) Please complete the following questions.

(a) What is the contour interval on this map segment? _____

(b) What is the approximate elevation at Lawton Ranch, near the mouth of the canyon? _____;

and, what is the elevation in the northwest portion of the map segment? _____

Given these two elevations, roughly determine the local relief of the alluvial fan and valley. _____

(c) What processes produce an alluvial fan? Why do these not occur in humid regions? _____

_____

_____

6. Complete the labels and leader lines for Figure 15.23c that portray a **bolson** in a desert landscape.

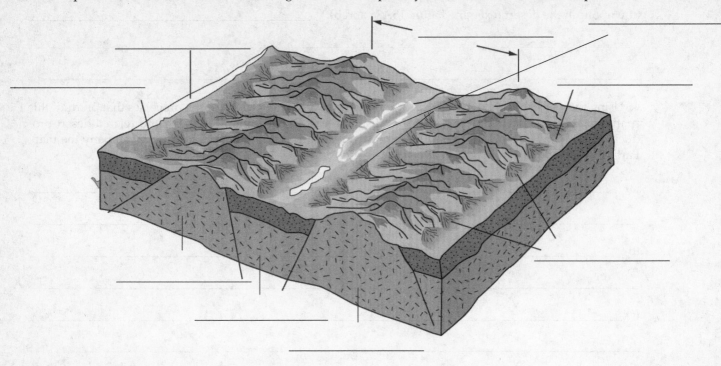

✳ STEP 4: The Colorado River System

1. Construct a simple bar graph depicting the Colorado River discharge at Lees Ferry for the years 1917, 1934, 1977, and 1984, and the 1930 through 2003 average flow of the river (see the data given in Focus Study 15.1, Focus Study Figure 15.1.1j and Table 1 "Estimated Colorado River Budget through 2003").

   Use the years noted along the bottom axis to begin the five bar graphs. On the left-side axis, design, determine, and add an appropriate scale of flow-discharge amounts (in acre-feet per year) to use in plotting the height of each bar graph.

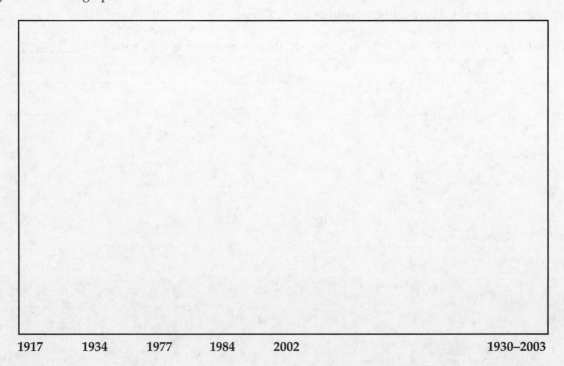

| 1917 | 1934 | 1977 | 1984 | 2002 | | 1930–2003 |

2. What is the total Colorado River water allocation for the seven U.S. states and Mexico? What was the total average flow, as recorded at Lees Ferry for 2002? 2003? _____

3. What is the annual loss of water from Lake Powell from bank storage? What are the geologic factors that contribute to this water loss? What is the annual evaporation water loss from Lake Powell? What is the loss of water due to phreatophytes? How does the total of these three sources compare to Mexico's annual water allocation? _____

_____

_____

_____

4. What method was used to allocate the water from the Colorado River? What was the number of years of records used in their analysis? Do you think this was a sufficiently large data set? _____

_____

_____

5. What is your evaluation of the procedure used to allocate Colorado River water? What assumptions turned out to be incorrect and what factors did they overlook? Were there later actions that made the situation better? Were there actions that made the situation worse? What effect did building the Glen Canyon Dam have on the overall system? _____

_____

_____

_____

6. What do you think the effects of a warmer, drier climate and a larger population will be on demands on the river? What actions are being taken to remedy the situation? What actions would you recommend?

_____

_____

_____

_____

_____

✳ STEP 5: NetWork—Internet Connection

There are many Internet addresses (URLs) listed in this chapter of the *Geosystems* textbook. Go to any two URLs, or "Destination" links on the *Geosystems* Home Page, and briefly describe what you find.

1. _____ : _____

_____

2. _____ : _____

_____

SAMPLE SELF-TEST

(Answers appear at the end of the study guide.)

1. Dry, arid climates occupy approximately what percentage of Earth's land surface?

   a. 55%
   b. 44%
   c. 26%, and 35% including all semi-arid lands
   d. 12%

2. The term eolian refers to

   a. stream-related processes
   b. sand dune fields only
   c. wind-eroded, wind-transported, and wind-deposited
   d. weathering and mass movement in humid regions

3. The formation of desert pavement

   a. involves both wind and water
   b. is produced by abrasion only
   c. is not related to deflation
   d. is greatly restricted in its occurrence on Earth

4. Nature's version of sandblasting is called

   a. deflation
   b. eolian deposition
   c. abrasion
   d. saltation

5. The Grand Ar Rub' al Khali is an example of a

   a. gibber plain
   b. reg desert
   c. sand sea
   d. stream-eroded area in the Middle East

6. Relative to loess deposits,

    a. they are principally composed of sands and gravels
    b. they are formed of fine-grained clays and silts
    c. occurrences are found only in the United States
    d. they form only following glacial activity

7. Which of the following correctly describes a dry stream bed that is intermittently filled with water?

    a. wash
    b. wadi
    c. arroyo
    d. all of these are correct

8. An exotic stream
    a. is any river that flows from an arid region to one of adequate precipitation
    b. is one involved in a particular historical event
    c. is one whose entire course is in a desert region
    d. is any river that arises in a humid region and then passes through to an arid region

9. Relative to the discharge of the Colorado River system, the construction of a large multi-purpose dam such as Glen Canyon augments (increases) streamflows and helps sustain the available water in the river.

    a. true
    b. false

10. The flood control measures and facilities on the Colorado River worked to prevent flooding during the spring of 1983.

    a. true
    b. false

11. The Qattara Depression in Egypt is a product of deflation.

    a. true
    b. false

12. Loess deposits form loose, gradual slopes, unstable for construction or road cuts.

    a. true
    b. false

13. The smallest features produced by saltation are called ripples.

    a. true
    b. false

14. Loess deposits in China exceed 300 m (984 ft).

    a. true
    b. false

15. Desert regions are characterized by high potential evapotranspiration, low precipitation, high evaporation, high input of insolation, and high radiative heat losses at night.

    a. true
    b. false

16. Some desert plants depend on the rushing, crashing waters in a wash to germinate.

    a. true
    b. false

17. John Wesley Powell specifically called on the federal government to build large multi-purpose dams and reservoirs in the arid west.

    a. true
    b. false

18. The Basin and Range Province of the western United States is an example of a horst and graben landscape.

    a. true
    b. false

16

THE OCEANS, COASTAL PROCESSES, AND LANDFORMS

Coastal regions are unique and dynamic environments. Most of Earth's coastlines are relatively new and are the setting for continuous change. The land, ocean, and atmosphere interact to produce tides, waves, erosional features, and depositional features along the continental margins. The interaction of vast oceanic and atmospheric masses is dramatic along a shoreline. At times, the ocean attacks the coast in a stormy rage of erosive power; at other times, the moist sea breeze, salty mist, and repetitive motion of the water are gentle and calming. All these coastal processes are being magnified in this era of a rising sea level.

As you read Chapter 16 and learn of the littoral zone, remember the role of humans and human impacts on the coastline. The chapter builds to a discussion of certain human impacts, a specific case study of Marco Island, Florida, and a focus study discussing an environmental approach to shoreline planning.

OUTLINE HEADINGS AND KEY TERMS

The first-, second-, and third-order headings that divide Chapter 16 serve as an outline for your notes and studies. The key terms and concepts that appear **boldface** in the text are listed here under their appropriate heading in *bold italics*. All these highlighted terms appear in the text glossary. Note the check-off box (❏) so you can mark your progress as you master each concept. These terms should be in your reading notes or used to prepare note cards. The ⊘ icon indicates that there is an accompanying animation on the Student CD. The ✹ icon indicates that there is an accompanying satellite or notebook animation on the Student CD.

The outline headings and terms for Chapter 16:

Global Oceans and Seas

⊘ **Midlatitude Productivity**

Chemical Composition of Seawater

   ❏ *salinity*

Ocean Chemistry

Average Salinity: 35‰

   ❏ *brine*
   ❏ *brackish*

Physical Structure of the Ocean

Coastal System Components

Inputs to the Coastal System

The Coastal Environment and Sea Level

   ❏ *littoral zone*
   ❏ *mean sea level (MSL)*

Coastal System Actions

⊘ **Monthly Tidal Cycle**

⊘ **Wave Motion/Wave Refraction**

⊘ **Beach Drift, Coastal Erosion**

Tides

   ❏ *tides*

Causes of Tides

   ❏ *flood tides*
   ❏ *ebb tides*

Spring and Neap Tides

   ❏ *spring tides*
   ❏ *neap tide*

Tidal Power

Waves

- ❑ *waves*
- ❑ *swells*
- ❑ *breaker*

Wave Refraction

- ❑ *wave refraction*
- ❑ *longshore current*
- ❑ *beach drift*

Tsunami, or Seismic Sea Wave

- ❑ *tsunami*

Coastal System Outputs

❂ **Coastal Stabilization Structures**

Erosional Coastal Processes and Landforms

- ❑ *wave-cut platform*

Depositional Coastal Processes and Landforms

- ❑ *barrier spit*
- ❑ *bay barrier*
- ❑ *lagoon*
- ❑ *tombolo*

Beaches

- ❑ *beach*

Maintaining Beaches

Barrier Formations

❀ **Hurricane Isabel Satellite Loop**

❀ **Hurricane Georges Satellite Loop**

- ❑ *barrier beaches*
- ❑ *barrier islands*

Barrier Island Origin and Hazards

Biological Processes: Coral Formations

- ❑ *coral*

Coral Reefs

Coral Bleaching

Wetlands, Salt Marshes, and Mangrove Swamps

- ❑ *wetlands*

Coastal Wetlands

- ❑ *salt marshes*
- ❑ *mangrove swamps*

Marco Island

Human Impact on Coastal Environments

SUMMARY AND REVIEW

News Reports, Focus Study, and High Latitude Connection

News Report 16.1: Sea-Level Variations and the Present MSL Increase

News Report 16.2: Engineers Nourish a Beach

Focus Study 16.1: An Environmental Approach to Shoreline Planning

High Latitude Connection 16.1: A Rebounding Shoreline and Coastal Features

---

URLs listed in Chapter 16

International Year of the Ocean:
http://www.yoto98.noaa.gov/

World Resources Institute ocean study:
http://www.wri.org/wri/sdis/maps/coasts/cri_idx.html

National Ocean Service:
http://www.nos.noaa.gov/

Global Sea-Level Observing Service:
http://www.pol.ac.uk/psmsl/gb.html
http://www.pol.ac.uk/psmsl/programmes/

TOPEX/Poseidon **satellite:**
http://topex-www.jpl.nasa.gov/

Scripps Institution of Oceanography library:
http://scilib.ucsd.edu/sio/tide/

Tsunami links:
http://www.pmel.noaa.gov/tsunami/
http://www.geophys.washington.edu/tsunami/welcome.html

Beach nourishment, Duke University:
http://www.env.duke.edu/psds/nourishment.htm

Cape Hatteras Lighthouse saving:
http://www.ncsu.edu/coast/chl/

Coral reefs:
http://www.wri.org/wri/indictrs/coastrsk.htm
http://www.coral.noaa.gov/gcrmn/
http://www.usgs.gov/coralreef.html

KEY LEARNING CONCEPTS FOR CHAPTER 16

The following key learning concepts help guide your reading and comprehension efforts. The operative word is in *italics*. Use these carefully to guide your reading of the chapter and note that STEP 1 asks you to work with these concepts. These same learning concepts are used in organizing the summary and review section at the end of the chapter—grouping together definitions, a list of key terms, and review questions.

After reading the chapter and using this study guide, you should be able to:

- *Describe* the chemical composition of seawater and the physical structure of the ocean.
- *Identify* the components of the coastal environment, and *list* the physical inputs to the coastal system, including tides and mean sea level.
- *Describe* wave motion at sea and near shore, and *explain* coastal straightening as a product of wave refraction.
- *Identify* characteristic coastal erosional and depositional landforms.
- *Describe* barrier islands and their hazards as they relate to human settlement.
- *Assess* living coastal environments: corals, wetlands, salt marshes, and mangroves.
- *Construct* an environmentally sensitive model for settlement and land use along the coast.

✳ STEP 1: Critical Thinking Process

Using your interest and learning, and the following questions as guidelines only, briefly discuss your experience with this chapter. In examining your learning you need not go through each of these questions in detail, simply provide an overview of your critical thinking process as it relates to some aspect of this chapter.

- What did you know about the learning concept before you began?
- Which information sources did you use in your learning (text, class, other)?
- Were you able to complete the action stated in the learning concept? What did you learn?
- Are there any aspects of the concept about which you want to know more?

Critical Thinking and Chapter 16: _____

_____

_____

_____

_____

✳ STEP 2: Global Oceans and Seas

1. Describe the chemical composition of seawater (principal dissolved solids and average salinity). _____

_____

_____

2. State the ocean's average salinity using the five expressions given in the text: _____

_____

_____

3. Contrast brine and brackish seawater and give examples of each. _____

_____

_____

4. Identify the sample of 27 oceans and seas indicated by number in Figure 16.2. Write the correct name in the spaces provided below <u>and</u> the country or continent it is nearest.

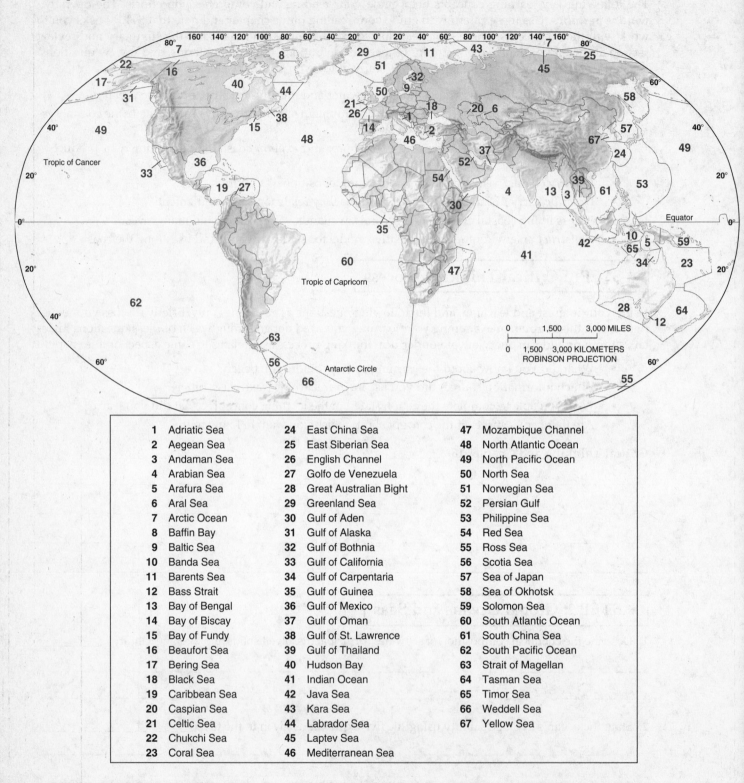

1	Adriatic Sea	24	East China Sea	47	Mozambique Channel
2	Aegean Sea	25	East Siberian Sea	48	North Atlantic Ocean
3	Andaman Sea	26	English Channel	49	North Pacific Ocean
4	Arabian Sea	27	Golfo de Venezuela	50	North Sea
5	Arafura Sea	28	Great Australian Bight	51	Norwegian Sea
6	Aral Sea	29	Greenland Sea	52	Persian Gulf
7	Arctic Ocean	30	Gulf of Aden	53	Philippine Sea
8	Baffin Bay	31	Gulf of Alaska	54	Red Sea
9	Baltic Sea	32	Gulf of Bothnia	55	Ross Sea
10	Banda Sea	33	Gulf of California	56	Scotia Sea
11	Barents Sea	34	Gulf of Carpentaria	57	Sea of Japan
12	Bass Strait	35	Gulf of Guinea	58	Sea of Okhotsk
13	Bay of Bengal	36	Gulf of Mexico	59	Solomon Sea
14	Bay of Biscay	37	Gulf of Oman	60	South Atlantic Ocean
15	Bay of Fundy	38	Gulf of St. Lawrence	61	South China Sea
16	Beaufort Sea	39	Gulf of Thailand	62	South Pacific Ocean
17	Bering Sea	40	Hudson Bay	63	Strait of Magellan
18	Black Sea	41	Indian Ocean	64	Tasman Sea
19	Caribbean Sea	42	Java Sea	65	Timor Sea
20	Caspian Sea	43	Kara Sea	66	Weddell Sea
21	Celtic Sea	44	Labrador Sea	67	Yellow Sea
22	Chukchi Sea	45	Laptev Sea		
23	Coral Sea	46	Mediterranean Sea		

2.	26.	38.
7.	31.	43.
11.	32.	44.
14.	33.	46.
23.	36.	48.
50.	54.	63.
51.	57.	64.
52.	59.	66.
53.	60.	67.

5. Describe the physical structure (percent of the oceanic mass and temperature) of the following components.

(a) Mixing zone: _____

(b) Thermocline transition zone: _____

(c) Deep cold zone: _____

✳ STEP 3: Coastal Systems

1. List and briefly describe the physical components (inputs) to the coastal system as discussed in the text.

_____

_____

_____

_____

_____

2. Using the following illustration derived from Figure 16.5, label the elements noted in the littoral zone.

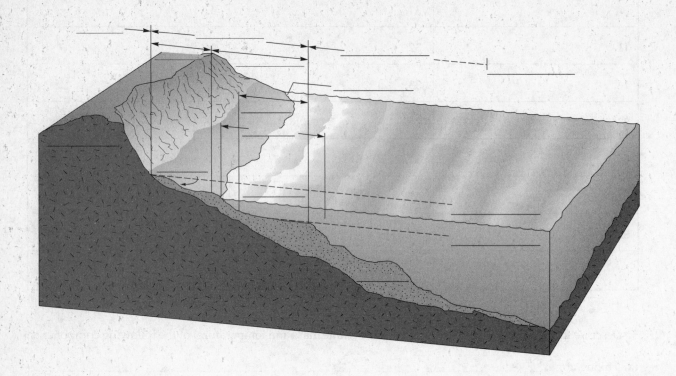

3. In Figure 16.6, how was a satellite able to determine sea level? How precise were the *TOPEX/Poseidon* measurements? What is meant by sea-surface topography? _____

_____

_____

_____

4. What is mean sea level? _____

_____

5. Relative to sea level and News Report 16.1:

(a) How does sea level vary along the U.S. coastline? _____

_____

_____

(b) According to the text, what changes have occurred in sea level?

Since 18,000 B.P.: _____

In the last 100 years: _____

(c) Briefly describe the forecast for future sea levels.

• Is the present rate of rise faster than observed previously? By how much? _____

_____

• Worst-case sea-level rise in the next 100 years: _____

• Most probable sea-level rise in the next 100 years: _____

(d) How do ocean temperatures relate to sea level heights? Relate the ocean temperature measurements gathered off the southern California coastline. Explain. _____

_____

_____

6. Draw a diagram showing the astronomical relationships that occur during the following times (Figure 16.7):

(a) spring tide

(b) neap tide

7. Are there any tidal-electrical generating stations operating presently in North America? How many are operating in North America? Worldwide? How do they generate electricity? _____

_____

_____

_____

8. Using the text and Figure 16.9, differentiate between *waves of transition* and *waves of translation*, include the physical conditions that are involved in each phase of wave action.

(a) transition: _____

_____

(b) translation: _____

_____

9. What is a tsunami? How are they formed? Describe their speed and height as they travel in the open ocean and when they approach the coast. _____

_____

_____

_____

10. How are tsunamis detected? Describe the North American warning systems in place. Why was there no warning for the December 26, 2004, tsunami in the Indian Ocean? What areas could potentially generate tsunamis that would affect North America? _____

_____

_____

_____

✳ STEP 4: Coastal System Outputs

1. Complete the appropriate labeling for an erosional coastline using this illustration derived from Figure 16.13.

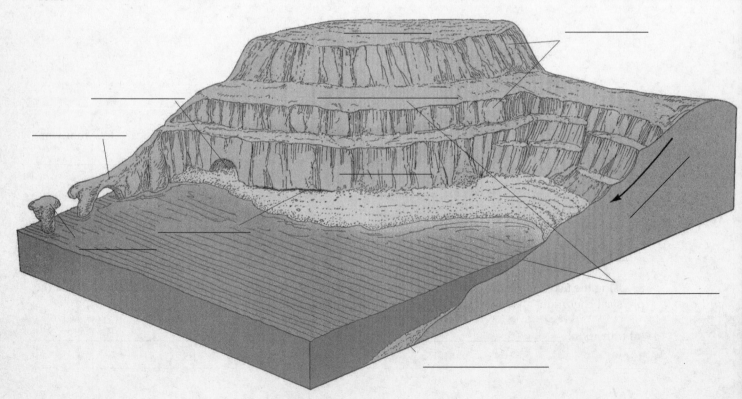

(a) Where is this type of erosional coastline found on Earth? Give a few examples beginning with the photo inserts in Figure 16.13. _____

_____

_____

_____

(b) Describe what produced the landform pictured in Figure 16.13e. _____

_____

2. Complete the appropriate labeling for this depositional coastline derived from Figure 16.14.

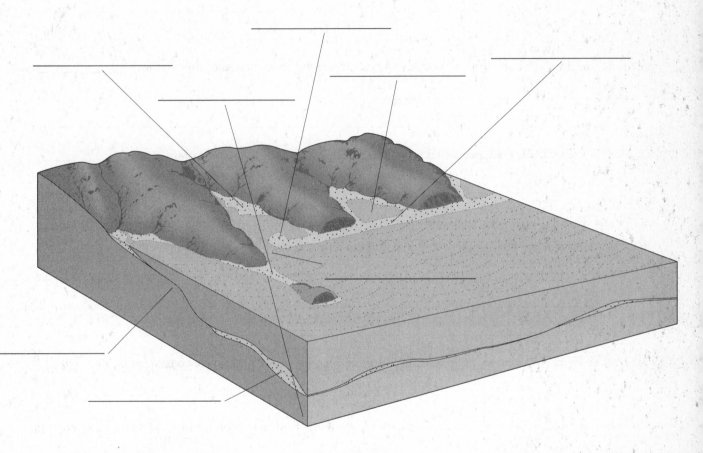

(a) Where is this type of depositional coastline found on Earth? Give a few examples beginning with the photo inserts in Figure 16.13a and b.

_____

_____

✳ STEP 5: Human Interactions

1. Societies use several specific physical constructions to intervene in littoral and beach drift and longshore, or littoral, currents. Explain the function of each and mention any consequences that might result from each construction.

(a) breakwater: _____

_____

(b) jetty: _____

_____

(c) groin: _____

_____

(d) In Figure 16.15b what effects do you see resulting from the construction of these jetties? _____

_____

_____

2. How is it possible for engineers to nourish(!) a beach (News Report 16.2)? Explain. _____

_____

3. Briefly describe what is pictured and mapped in Figure 16.18. What event occurred here? Analyze the differences you see between the older topo map and the newer aerial photograph composite. _____

_____

_____

_____

(a) How does this example relate to what is pictured in Figure 16.15c in terms of coastal processes? _____

_____

(b) What is your assessment of the hazards associated with construction on barrier islands or other coastal lowlands (see also Figure 16.25)? _____

_____

_____

_____

4. Relative to Earth's coral formations, what condition is emerging worldwide? Are there any clues that scientists are pointing to as a cause for this condition? _____

_____

_____

5. Relative to the following illustration derived from Focus Study 16.1, Figure 1, label and identify each of the components characteristic of the coastal environment along the New Jersey shore. Use the spaces provided for your labels and descriptions.

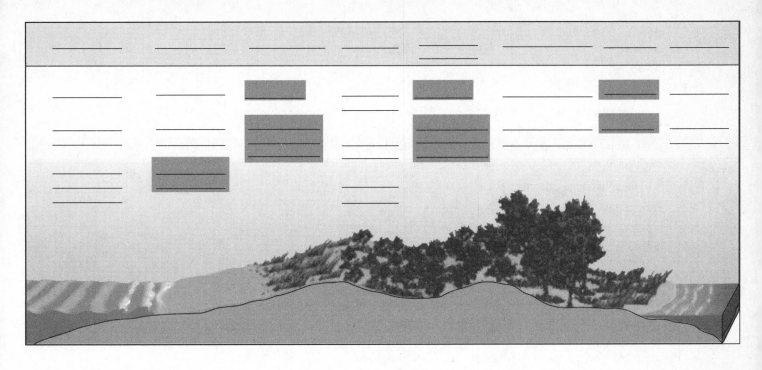

✳ STEP 6: NetWork—Internet Connection

There are many Internet addresses (URLs) listed in this chapter of the *Geosystems* textbook. Go to any two URLs, or "Destination" links on the *Geosystems* Home Page, and briefly describe what you find.

1. _____ : _____

_____

2. _____ : _____

_____

SAMPLE SELF-TEST
(Answers appear at the end of the study guide.)

1. Another general term for the coastal environment is

    a. shoreline
    b. coast
    c. coastline
    d. littoral zone

2. Relative to mean sea level,

    a. a consistent value has yet to be determined due to all the variables involved in producing the tides
    b. it is at a similar level along all of the North American coast
    c. it is calculated based on average tidal levels recorded hourly at a given site over a period of 19 years
    d. sea level along the Gulf Coast is the lowest for the coasts of the lower 48 states

3. Which of the following is <u>incorrect</u>?

    a. rising tides—flood tides
    b. ebb tides—falling tides
    c. spring tides—happen only in the spring
    d. spring tides—extremes of high and low tides

4. Waves travel in wave trains that

    a. are produced by storm centers and generating regions far distant from where they break
    b. usually form relatively close to the affected coastline
    c. are developed in relation to the specific contours and shape of the affected coastline
    d. are called waves of translation as they travel in the open sea

5. A tsunami is

    a. also known as a tidal wave
    b. also known as a seismic sea wave
    c. a large wave form that travels in the open sea roughly at the same velocity that is witnessed at the shore
    d. caused by a strong storm center in the open sea

6. Sand deposited in a long ridge extending out from a coast, connected to land at one end, is called a

    a. barrier spit
    b. lagoon
    c. bay barrier
    d. tombolo

7. The most extensive chain of barrier islands in the world is along

    a. the western coast of Australia
    b. the coasts that surround the periphery of the Indian Ocean
    c. the east coast of the Asian landmass
    d. the Atlantic and Gulf coasts of North America, extending some 5000 km

8. Embayments and ria coasts are characteristic of

    a. erosional coasts
    b. emergent coasts
    c. submergent coasts
    d. the western margin of North America

9. The coastal environment is generally called the littoral zone.

   a. true
   b. false

10. Earth's tidal bulges are only produced by Earth's centrifugal forces as opposed to gravitational forces.

    a. true
    b. false

11. Earth's tidal bulges are partially attributable to the pull of the Sun and the Moon.

    a. true
    b. false

12. Spring tides reflect greater tidal ranges, whereas neap tides produce lesser tidal ranges.

    a. true
    b. false

13. Electricity has yet to be produced by tidal power generation but remains a viable alternative for the future.

    a. true
    b. false

14. A wave-cut platform, or wave-cut terrace, is an example of a coastal depositional feature.

    a. true
    b. false

15. Beaches are dominated by sand because quartz (SiO_2) is so abundant and resists weathering.

    a. true
    b. false

16. The barrier islands off South Carolina were struck for the <u>first</u> time by Hurricane Hugo in 1989, with the local residents never having experienced such storms before during this century.

    a. true
    b. false

17. Corals are not capable of building major landforms; rather, they form on the top of already existing landmasses.

    a. true
    b. false

18. A salt marsh is not a wetland, whereas a mangrove swamp is a wetland.

    a. true
    b. false

19. Relative to possible development, the front, or primary dune, is _____;

    the trough behind the primary dune is _____;

    the secondary dune is _____;

    the backdune is _____;

    and the bayshore and bay is _____

20. The circular undulations in the open sea are known as waves of _____,

whereas the wave action near shore forms breakers and waves of _____

21. A _____

consists of sand deposited in a long ridge extending out from a coast; it partially crosses and blocks the mouth of a

bay. This spit becomes a _____

_____

if it completely cuts off the bay from the ocean. When sand deposits connect the shoreline with an offshore island or

sea stack it is a _____

22. Relative to coastal wetlands, _____ tend to form north of the 30th parallel, whereas

_____ form equatorward of that point.

17

GLACIAL AND PERIGLACIAL PROCESSES AND LANDFORMS

As the text states, a large measure of the freshwater on Earth is frozen, with the bulk of that ice sitting restlessly in just two places—Greenland and Antarctica. The remaining ice covers various mountains and fills alpine valleys. More than 29 million km^3 (7 million mi^3) of water is tied up as ice, or about 77% of all freshwater. Distinctive landscapes are produced by the ebb and flow of glacial ice masses. Other regions experience permanently or seasonally frozen ground that affects the topography and human habitation.

These deposits of ice, laid down over several million years, provide an extensive frozen record of Earth's climatic history and perhaps some clues to its climatic future. The inference is that rather than being static and motionless, distant, frozen places, these frozen lands are dynamic and susceptible to change, just as they have been over Earth's history. Changes in the ice mass worldwide signal vast climatic change and further glacio-eustatic increases in sea level. As you study this chapter, keep this significance in mind.

As an example, please consult Mark F. Meier (U.S. Geological Survey, Tacoma, Washington), "Contribution of Small Glaciers to Global Sea Level," *Science* Vol. 266 (21 December 1984): 1416–1421. He states in his summary that,

> Observed long-term changes in glacier volume and hydrometeorological mass balance models yield data on the transfer of water from glaciers, excluding those in Greenland and Antarctica, to the oceans. These glaciers appear to account for a third to half of observed rise in sea level, approximately that fraction not explained by thermal expansion of the ocean.

A loss of polar ice mass, augmented by melting of alpine and mountain glaciers (which experienced more than 30% decrease in overall ice mass during the last century), will affect sea-level rise. Worldwide, glacial ice is in retreat, melting at rates exceeding anything in the ice record. In the European Alps alone some 75% of the glaciers have receded in the past 50 years, losing more than 50% of their ice mass since 1850. At this rate, the European Alps will have only 20% of their preindustrial glacial ice left by 2050. The IPCC assessment states that "there is conclusive evidence for a worldwide recession of mountain glaciers. This is among the clearest and best evidence for a change in energy balance at the Earth's surface since the end of the 19th century."

OUTLINE HEADINGS AND KEY TERMS

The first-, second-, and third-order headings that divide Chapter 17 serve as an outline for your notes and studies. The key terms and concepts that appear **boldface** in the text are listed here under their appropriate heading in ***bold italics***. All these highlighted terms appear in the text glossary. Note the check-off box (❑) so you can mark your progress as you master each concept. These terms should be in your reading notes or used to prepare note cards. The ✪ icon indicates that there is an accompanying animation on the Student CD. The ✸ icon indicates that there is an accompanying satellite or notebook animation on the Student CD.

The outline headings and terms for Chapter 17:

Rivers of Ice
    ❑ *glacier*
    ❑ *snowline*

Alpine Glaciers

- ❏ *alpine glacier*
- ❏ *cirque*
- ❏ *icebergs*

Continental Glaciers

- ❏ *continental glacier*
- ❏ *ice sheet*
- ❏ *ice cap*
- ❏ *ice field*

Glacial Processes

✺ **Budget of a Glacier, Mass Balance**
✺ **Flow of Ice within a Glacier**

Formation of Glacial Ice

- ❏ *firn*
- ❏ *glacial ice*

Glacial Mass Balance

- ❏ *firn line*
- ❏ *ablation*

Glacial Movement

- ❏ *crevasses*

Glacier Surges

- ❏ *glacier surge*

Glacial Erosion

- ❏ *abrasion*

Glacial Landforms

✹ **Teton Glacier Notebook**

Erosional Landforms Created by Alpine Glaciation

- ❏ *arêtes*
- ❏ *col*
- ❏ *horn*
- ❏ *bergschrund*
- ❏ *tarns*
- ❏ *paternoster lakes*
- ❏ *fjord*

Depositional Landforms Created by Alpine Glaciation

Glacial Drift

- ❏ *glacial drift*
- ❏ *stratified drift*
- ❏ *till*

Moraines

- ❏ *moraine*
- ❏ *lateral moraine*
- ❏ *medial moraine*
- ❏ *terminal moraine*

Erosional and Depositional Landforms Created by Continental Glaciation

- ❏ *till plain*
- ❏ *outwash plain*
- ❏ *esker*
- ❏ *kettle*
- ❏ *kame*
- ❏ *roche moutonnée*
- ❏ *drumlin*

Periglacial Landscapes

- ❏ *periglacial*

Geography of Permafrost

- ❏ *permafrost*

Continuous and Discontinuous Zones

Behavior of Permafrost

- ❏ *active layer*

Ground Ice and Frozen Ground Phenomena

Frost Action Processes

Frost Action Landforms

- ❏ *patterned ground*

Hillslope Processes: Gelifluction and Solifluction

- ❏ *solifluction*
- ❏ *gelifluction*

Thermokarst Landscapes

Humans and Periglacial Landscapes

The Pleistocene Ice-Age Epoch

- ❏ *ice age*

Changes in the Landscape

Lowered Sea Levels and Lower Temperatures

Paleolakes

- ❏ *paleolakes*

Deciphering Past Climates: Paleoclimatology

KEY LEARNING CONCEPTS FOR CHAPTER 17

The following key learning concepts help guide your reading and comprehension efforts. The operative word is in *italics*. Use these carefully to guide your reading of the chapter and note that STEP 1 asks you to work with these concepts. These same learning concepts are used in organizing the summary and review section at the end of the chapter—grouping together definitions, a list of key terms, and review questions.

    After reading the chapter and using this study guide, you should be able to:

- *Differentiate* between alpine and continental glaciers, and *describe* their principal features.
- *Describe* the process of glacial ice formation, and *portray* the mechanics of glacial movement.
- *Describe* characteristic erosional and depositional landforms created by alpine glaciation and continental glaciation.
- *Analyze* the spatial distribution of periglacial processes, and *describe* several unique landforms and topographic features related to permafrost and frozen ground phenomena.
- *Explain* the Pleistocene ice-age epoch and related glacials and interglacials, and *describe* some of the methods used to study paleoclimatology.

❋ STEP 1: Critical Thinking Process

Using your interest and learning, and the following questions as guidelines <u>only</u>, briefly discuss your experience with this chapter. In examining your learning you need not go through each of these questions in detail, simply provide an overview of your critical thinking process as it relates to some aspect of this chapter.

- What did you know about the learning concept before you began?
- Which information sources did you use in your learning (text, class, other)?
- Were you able to complete the action stated in the learning concept? What did you learn?
- Are there any aspects of the concept about which you want to know more?

Critical Thinking and Chapter 17: _____

_____

_____

_____

_____

❋ STEP 2: Rivers of Ice

1. In the high-altitude photo and inset map of the Mount McKinley region shown in Figure 17.2, how many alpine glaciers or valley glaciers can you identify?

_____

_____

2. Name a specific example of each of the following as presented in the text, the chapter opener figure on pages 530–531, and Figures 17.1, 17.2, 17.3, 17.4, 17.5,17.6, and 17.8c:

(a) alpine glacier: _____

(b) piedmont glacier: _____

(c) ice field: _____

(d) ice cap: _____

(e) ice sheets (name 2): _____

(f) fjords: _____

(g) crevasses: _____

(h) nunataks: _____

3. Differentiate between a *continental glacier* and an *alpine glacier*.

_____

_____

4. In October 1996, the _____ volcano beneath the_____Ice

Cap in Iceland began erupting. Figure 17.5a is an image of this ice cap (note labels on the locator map). Massive floods are created by such eruptions beneath the ice. In Iceland what do they call the path of floodwaters to the ocean? _____

✻ STEP 3: Glacial Processes

1. Briefly describe the formation process that produces glacial ice from snow.

_____

_____

_____

2. What is a glacier's mass balance (Figure 17.7b)?

_____

_____

3. Complete the labels on the following mass balance diagram.

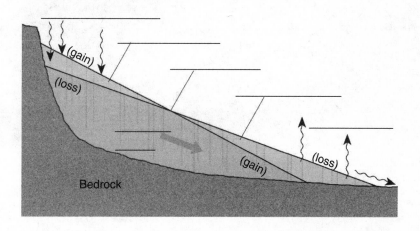

4. Relative to the South Cascade glacier (News Report 17.1), what is the trend in its mass balance between 1955 and 2002 (see the graph in Figure 17.1.1)? Assess the change that has occurred and describe.

_____

_____

_____

5. Describe what is meant by *glacial surge* and give at least two examples of this type of glacial movement.

_____

_____

_____

6. Complete the labels on the following illustration of a typical retreating alpine glacier derived from Figure 17.7a.

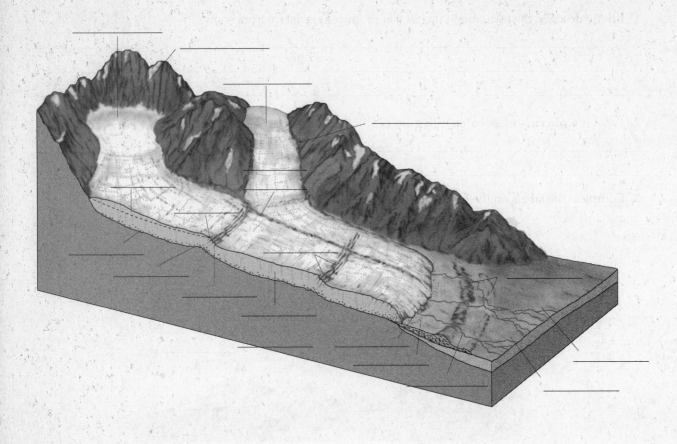

✳ STEP 4: Glacial Landforms

1. Using Figure 17.10a, b, and c, the eight inset photos, and the text, characterize each of the stages of a landscape subjected to fluvial (a) and glacial processes and actions during (b) and after (c) glaciation. Detail a few features that are indicative of each phase.

(a) Before: _____

_____

(b) During: _____

_____

(c) After: _____

_____

2. Describe the glacial features you have learned that appear in Figure 17.11a: _____

_____

_____

_____

3. What is the feature shown in Figure 17.12? How was it formed? (Also, revisit Figure 17.1b.)

_____

_____

_____

4. List the features common to alpine (valley) glaciation that are <u>absent</u> in continental glaciation (see Table 17.1).

_____

_____

_____

5. Compare and contrast the following two landforms produced by continental glaciation.

(a) roche moutonnée: _____

_____

(b) Does this description fit the photograph in Figure 17.16a of Lembert Dome? Explain. _____

_____

(c) drumlin: _____

_____

_____

(d) Can you identify these features on the topographic map in Figure17.17a? _____

_____

_____

✳ STEP 5: Periglacial Landscapes

1. Define each and distinguish among the following terms:

(a) permafrost: _____

_____

(b) ground ice: _____

_____

(c) active layer: _____

_____

2. Using Figure 17.19 as a reference, what impact do you think the record temperatures in Canada are having on the periglacial environment? Portions of Canada were 2 C° to 5 C° above average in 1998 and temperatures reached over 30°C (86°F) as far north as James Bay during the spring 1999.

_____

_____

_____

3. List two examples of ground ice formations or landforms.

(a) _____

(b) _____

4. Describe what is depicted in Figure 17.24. Why is the cabin failing? The construction of "Utilidors" shown in Figure 17.25a is necessary because of what conditions? Relate these in your answer. _____

_____

_____

✳ STEP 6: Past Glaciation and Paleoclimatology

1. Define:

(a) ice age: _____

_____

(b) paleoclimatology: _____

_____

(c) Pleistocene: _____

_____

2. Relative to the text discussion of glacial and interglacial stages and Figure 17.26, describe the following:

(a) Illinoian (number of stages?): _____

_____

(b) Wisconsinan (number of stages?): _____

_____

(c) Sangamon: _____

_____

(d) Stage 23: _____

_____

3. In terms of paleoclimatology and past climatic changes over the long span of geologic history, what five physical factors are mentioned in the text as probable causes?

(a) _____

(b) _____

(c) _____

(d) _____

(e) _____

4. What are the three astronomical factors that may affect climatic cycles? What are their time periods?

(a) _____

(b) _____

(c) _____

5. Who studied the relationship between climate and celestial relations? When did he do his work? Were his ideas generally accepted by the scientific community at the time of his death? Is there evidence that supports his ideas? Which other scientist that you have studied had his ideas accepted only after his death?

_____

_____

_____

_____

_____

6. What do ice cores in Greenland have to do with deciphering past climates (Focus Study 17.1 and Figure 17.30 a, b, and c)? Why is the core from MIS-11 especially interesting? Explain using some specifics. _____

_____

_____

_____

_____

(a) Medieval Warm Period: _____

_____

(b) Little Ice Age: _____

_____

7. After reading Focus Study 17.1 about GRIP, GISP-2, and Dome-C, answer the following questions:

(a) Where are the drilling sites for GRIP, GISP-2, and Dome-C? _____

_____

(b) What is the maximum depth drilled at GRIP, GISP-2, and Dome-C? How far back in time does this represent for each?

GRIP: _____

GISP-2: _____

Dome-C: _____

8. What are the three reasons the ice age concept is currently being debated so intensely?

(a) _____

_____

(b) _____

_____

(c) _____

_____

9. After reviewing the High Latitude Connections 1.1, 5.1, 10.1, 11.1, 17.1, 21.1, and News Reports 17.1 and 17.2, briefly summarize what you have learned about the evidence and effects of climate change in polar regions. Do you think these represent clear evidence of climate change? Why or why not? Which do you find most dramatic and why do you find it so?

_____

_____

_____

_____

_____

_____

❄ STEP 7: Polar Regions

1. Be sure to examine the recent photo of the Amundsen-Scott Base at the South Pole in Figure 17.35. Read the caption and identify the structures.

2. Describe the physical criteria used to delimit the polar regions and briefly characterize each.

(a) Arctic: _____

_____

_____

(b) Antarctic: _____

_____

_____

3. Is there any evidence for an Arctic ice sheet during the last ice age (News Report 17.3)?

_____

_____

❄ STEP 8: NetWork—Internet Connection

There are many Internet addresses (URLs) listed in this chapter of the *Geosystems* textbook. Go to any two URLs, or "Destination" links on the *Geosystems* Home Page, and briefly describe what you find.

1. _____: _____

_____

2. _____: _____

_____

SAMPLE SELF-TEST
(Answers appear at the end of the study guide.)

1. A general term for a mass of perennial ice, resting on land or floating shelflike in the sea adjacent to land, is a/an

    a. snowline
    b. iceberg
    c. glacier
    d. ice field

2. Alpine glaciers include all of the following except a/an

    a. ice cap
    b. mountain glacier
    c. cirque glacier
    d. valley glacier

3. Relative to glacial mass balance, which of the following is incorrect?

   a. a positive net balance or negative net balance occurs during a cold period and warm period respectively
   b. glacial mass is reduced by evaporation, sublimation, and deflation
   c. glacial mass is reduced by a combination of processes called ablation
   d. the zone in the glacier where accumulation gains and losses begin is the equilibrium line
   e. worldwide, most glaciers are showing marked increases in mass at present

4. Glacial ice is

   a. essentially the same as snow
   b. formed after a long process that may take 1000 years in Antarctica
   c. also known as firn
   d. generally less dense than snow and firn

5. Glacial erosion specifically involves

   a. ablation
   b. deflation
   c. abrasion and plucking
   d. eskers and kames

6. Periglacial processes

   a. affect approximately 20% of Earth's land surface
   b. occur at high latitudes and high altitudes
   c. involve permafrost, frost action, and ground ice
   d. all of these are correct

7. Which of the following is incorrect relative to the Pleistocene ice age epoch?

   a. it began 1.65 million years ago
   b. it produced ice sheets and glaciers that covered 30 percent of Earth's land area
   c. at least 18 expansions of ice occurred over Europe and North America
   d. it represents a single continuous cold spell

8. Geomorphic landforms produced by valley glaciers include which of the following?

   a. drumlins
   b. eskers
   c. ice caps and sheets
   d. horns and cols

9. In terms of paleoclimatology, which of the following was not mentioned as a probable cause of past ice ages?

   a. the weight of the ice
   b. galactic and Earth–Sun relationships
   c. geophysical factors
   d. geographical–geological factors

10. Milankovitch proposed astronomical factors as a basic cause of ice ages.

    a. true
    b. false

11. The Great Salt Lake, a remnant of a paleolake (pluvial lake), reached record historic levels during the 1980s.

    a. true
    b. false

12. The definition of the Antarctic region is simply the extent of the Antarctic continental landmass.

    a. true
    b. false

13. Greenland actually contains more ice than does Antarctica at the present time.

    a. true
    b. false

14. A valley glacier is commonly associated with unconfined bodies of ice.

    a. true
    b. false

15. The Vatnajökull glacier of Iceland is presented as an example of an ice cap.

    a. true
    b. false

16. Internally, a glacier can continue to move forward even though its lower terminus is in retreat.

    a. true
    b. false

17. A **V**-shaped valley is characteristic of glaciated valleys, whereas a **U**-shape is more characteristic of a stream-eroded valley.

    a. true
    b. false

18. An ice wedge is a form of ground ice in periglacial landscapes.

    a. true
    b. false

19. Glacial till is unsorted, whereas glacial outwash is sorted and stratified as one would expect stream-deposited materials to be.

    a. true
    b. false

20. The cirques where the valley glacier originated form cirque walls that wear away; when this happens, an _____ _____, or a sharp ridge that divides two cirque basins forms. When two eroding cirques reduce this to form a pass or saddlelike depression, the term _____ is used. A pyramidal peak called a _____ results when several cirque glaciers gouge an individual mountain summit from all sides. A small mountain lake, especially one that collects in a cirque basin behind risers of rock material, is called a _____. Small, circular, stair-stepped lakes in a series are called _____ because they look like a string of rosary (religious) beads forming in individual rock basins aligned down the course of a glaciated valley.

PART FOUR:
Soils, Ecosystems, and Biomes

OVERVIEW—PART FOUR

Earth is the home of the only known biosphere in the Solar System: a unique, complex, and interactive system of abiotic and biotic components working together to sustain a tremendous diversity of life. Thus we begin Part Four of *Geosystems* and an examination of the geography of the biosphere—soils, ecosystems, and terrestrial biomes. Recall the description of the biosphere given in Chapter 1:

> The intricate, interconnected web that links all organisms with their physical environment is the **biosphere**. Sometimes called the **ecosphere**, the biosphere is the area in which physical and chemical factors form the context of life. The biosphere exists in the overlap among the abiotic spheres, extending from the sea floor and even the upper layers of the crustal rock to about 8 km (5 mi) into the atmosphere. Life is sustainable within these natural limits. In turn, life processes have powerfully shaped the other three spheres through various interactive processes. The biosphere evolves, reorganizes itself at times, faces some extinctions, and manages to flourish. Earth's biosphere is the only one known in the Solar System; thus, life as we know it is unique to Earth.

> Part Four is a synthesis of the many physical systems covered throughout the text.

Name: _____   Class Section: _____

Date: _____   Score/Grade: _____

18

THE GEOGRAPHY OF SOILS

Earth's landscape is generally covered with soil. Soil is a dynamic natural body and comprises fine materials, in which plants grow, and which is composed of both mineral and organic matter. Because of their diverse nature, soils are a complex subject and pose a challenge for spatial analysis. This chapter presents an overview of the modern system of soil classification used in the United States. Coverage of the Canadian system is in Appendix B of the text.

Please find Table 18.4, and utilize it to simplify this geographic study of soils. The table summarizes the 12 soil orders of the Soil Taxonomy. The table includes general location and climate association, areal coverage estimate, and basic description.

Also, for most soil orders, you will find small locator maps included, along with a picture of that soil's profile. These locator maps are derived from the worldwide distribution map presented in Figure 18.9. This map was adapted and redrawn from one prepared by the World Soil Resources Staff of the Natural Resources Conservation Service and published in November 1998. A new soil order appears on this map and is discussed in the chapter—the Gelisols.

OUTLINE HEADINGS AND KEY TERMS

The first-, second-, and third-order headings that divide Chapter 18 serve as an outline for your notes and studies. The key terms and concepts that appear **boldface** in the text are listed here under their appropriate heading in *bold italics*. All these highlighted terms appear in the text glossary. Note the check-off box (❏) so you can mark your progress as you master each concept. These terms should be in your reading notes or used to prepare note cards. The ✹ icon indicates that there is an accompanying satellite or notebook animation on the Student CD.

The outline headings and terms for Chapter 18:

- ❏ *soil*
- ❏ *soil science*

Soil Characteristics

Soil Profiles

- ❏ *pedon*
- ❏ *polypedon*

Soil Horizons

- ❏ *soil horizon*
- ❏ *humus*
- ❏ *eluviation*
- ❏ *illuviation*
- ❏ *solum*

Soil Properties

✹ Soil Ion Exchange: Soil Particles and Soil Water

Soil Color

Soil Texture

- ❏ *loam*

Soil Structure

Soil Consistence

Soil Porosity

Soil Moisture

Soil Chemistry

- ❏ *soil colloids*
- ❏ *cation-exchange capacity (CEC)*
- ❏ *soil fertility*

Soil Acidity and Alkalinity

Soil Formation Factors and Management

Natural Factors

The Human Factor

Soil Classification

Soil Taxonomy

- ❏ *Soil Taxonomy*

Pedogenic Regimes

- ❏ *pedogenic regimes*
- ❏ *laterization*
- ❏ *salinization*
- ❏ *calcification*
- ❏ *podzolization*
- ❏ *gleization*

Diagnostic Soil Horizons

- ❏ *epipedon*
- ❏ *diagnostic subsurface horizon*

The 12 Soil Orders of the Soil Taxonomy

Oxisols

- ❏ *Oxisols*

Aridisols

- ❏ *Aridisols*

Mollisols

- ❏ *Mollisfols*

Alfisols

- ❏ *Alfisols*

Ultisols

- ❏ *Ultisols*

Spodosols

- ❏ *Spodosols*

Entisols

- ❏ *Entisols*

Inceptisols

- ❏ *Inceptisols*

Gelisols

- ❏ *Gelisols*

Andisols

- ❏ *Andisols*

Vertisols

- ❏ *Vertisols*

Histosols

- ❏ *Histosols*

SUMMARY AND REVIEW

News Reports and Focus Study

News Report 18.1: Soil Is Slipping Through Our Fingers

News Report 18.2: Drainage Tiles, But Where to Go?

Focus Study 18.1: Selenium Concentration in Western Soils

URLs listed in Chapter 18

U.S. Department of Agriculture, Natural Resources Conservation Service:
http://www.nrcs.usda.gov/

Soil Science Society of America:
http://www.soils.org/

National Soil Survey Center:
http://soils.usda.gov

Natural Resources Conservation Service:
http://www.nrcs.usda.gov/

Agriculture Canada Soils sources:
http://sis.agr.gc.ca/cansis/intro.html
http://www.metla.fi/info/vlib/soils/old.htm

World soils:
http://www.nhq.nrcs.usda.gov/

KEY LEARNING CONCEPTS FOR CHAPTER 18

The following key learning concepts help guide your reading and comprehension efforts. The operative word is in *italics*. Use these carefully to guide your reading of the chapter and note that STEP 1 asks you to work with these concepts. These same learning concepts are used in organizing the summary and review section at the end of the chapter—grouping together definitions, a list of key terms, and review questions.

After reading the chapter and using this study guide, you should be able to:

- *Define* soil and soil science, and *describe* a pedon, polypedon, and typical soil profile.
- *Describe* soil properties of color, texture, structure, consistence, porosity, and soil moisture.
- *Explain* basic soil chemistry, including cation-exchange capacity, and *relate* these concepts to soil fertility.
- *Evaluate* the principal soil formation factors, including the human element.
- *Describe* the twelve soil orders of the Soil Taxonomy classification system, and *explain* their general occurrence.

✳ STEP 1: Critical Thinking Process

Using your interest and learning, and the following questions as guidelines <u>only</u>, briefly discuss your experience with this chapter. In examining your learning you need not go through each of these questions in detail, simply provide an overview of your critical thinking process as it relates to some aspect of this chapter.

- What did you know about the learning concept before you began?
- Which information sources did you use in your learning (text, class, other)?
- Were you able to complete the action stated in the learning concept? What did you learn?
- Are there any aspects of the concept about which you want to know more?

_____

_____

_____

_____

✳ STEP 2: Soil Characteristics and Properties

1. Define the horizons of a soil profile.

(a) O: _____

_____

(b) A: _____

_____

(c) E: _____

_____

(d) B: _____

_____

(e) C: _____

_____

(f) R: _____

_____

2. Label each specific horizon in the following illustration taken from Figure 18.1. As you identify them you may want to use colored pencils to color each horizon.

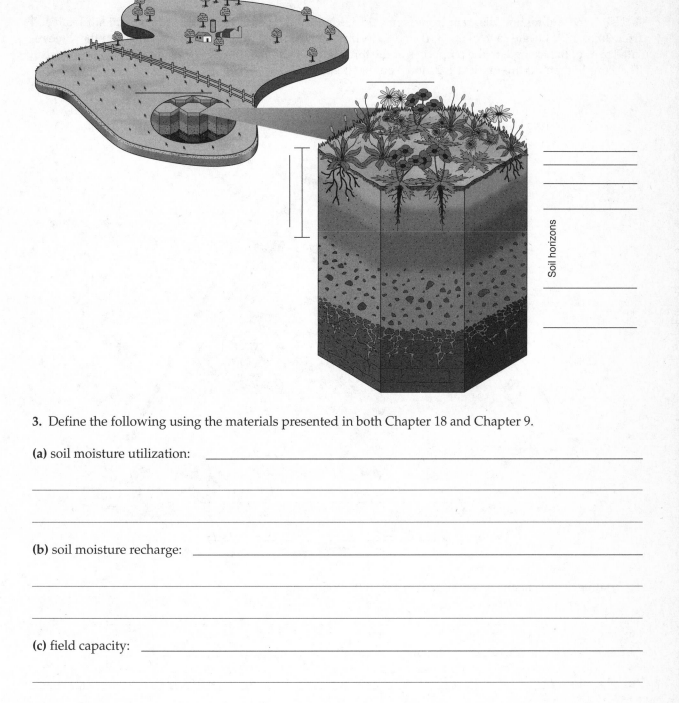

Soil horizons

3. Define the following using the materials presented in both Chapter 18 and Chapter 9.

(a) soil moisture utilization: _____

_____

_____

(b) soil moisture recharge: _____

_____

_____

(c) field capacity: _____

_____

_____

(d) wilting point: _____

_____

_____

(e) capillary (available) water: _____

_____

_____

4. Using the soil texture diagram from Figure 18.4, identify the textural analysis of a Miami silt loam soil from Table 18.1. Locate each of the A, B, and C horizons on the illustration and draw a line from the three related axes of the triangle to the properties of the three horizons so that these textural analyses lines intersect. The sampling points are marked 1, 2, and 3 on the illustration.

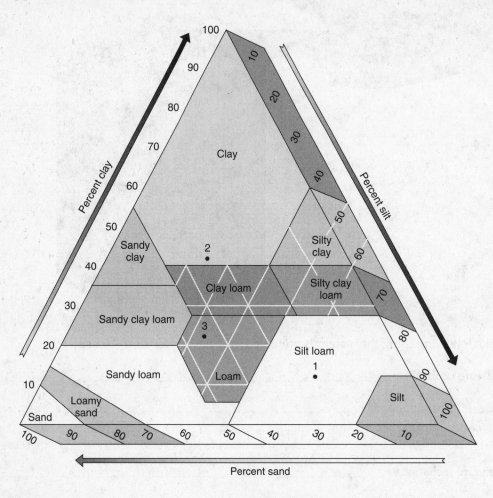

5. Relative to soil chemistry, characterize each of the following.

(a) pH: _____

_____

(b) alkalinity: _____

_____

(c) acidity: _____

_____

(d) cation-exchange capacity: _____

_____

✳ STEP 3: Soil Formation Factors and Management

1. List and briefly describe examples of the following non-human soil formation factors:

(a) Climate _____

_____

_____

(b) Vegetation, animal, and bacterial activity _____

_____

_____

(c) Topography _____

_____

_____

2. How many years are required to generate a few centimeters of soil? How long could it take to lose this amount of soil if poor practices are followed? List and describe some practices that encourage soil erosion.

_____

_____

_____

3. How many acres of farmland are lost in the U.S. each year due to mismanagement or conversion to non-agricultural uses? What percentage of U.S. and Canadian farmland is experiencing excessive rates of soil erosion? How much of the world's farmable land has been lost since 1950? How many new acres are lost each year?

_____

_____

_____

4. What are the economic costs of soil erosion in the U.S.? Worldwide? What are the costs of controlling erosion in the U.S.? Based on these numbers, what is the cost/benefit ratio of controlling soil erosion in the U.S.?

_____

_____

_____

5. What are the two main problems associated with irrigation of arid lands? Which soil order is usually found in these areas? Which parts of the world have these problems? What additional factor contributed to the problems at the Kesterson National Wildlife Refuge?

_____

_____

_____

_____

✳ STEP 4: Soil Classification

1. Using Table 18.4 and the world map in Figure 18.9, give a basic geographic description of representative locations for each order.

Oxisols: _____

_____

_____

Aridisols: _____

_____

_____

Mollisols: _____

_____

_____

Alfisols: _____

_____

_____

Ultisols: _____

_____

_____

Spodosols: _____

_____

_____

Entisols: _____

_____

_____

Inceptisols: _____

_____

_____

Gelisols: _____

_____

_____

Andisols: _____

_____

_____

Vertisols: _____

_____

_____

Histosols: _____

_____

_____

Optional from Appendix B:

2. Relative to the Canadian System of Soil Classification (CSSC): Using the table and Canadian soil map in Appendix B, give a basic geographic description of representative locations for each order.

Chernozemic: _____

_____

Solonetzic: _____

_____

Luvisolic: _____

_____

Podzolic: _____

_____

Brunisolic: _____

_____

Regosolic: _____

_____

Gleysolic: _____

_____

Organic: _____

_____

Crysolic: _____

_____

✳ STEP 5: NetWork—Internet Connection

There are many Internet addresses (URLs) listed in this chapter of the *Geosystems* textbook. Go to any two URLs, or "Destination" links on the *Geosystems* Home Page, and briefly describe what you find.

1. _____ : _____

_____

2. _____ : _____

_____

SAMPLE SELF-TEST
(Answers appear at the end of the study guide.)

1. Soil is

   a. a physical, geological product
   b. ground-up rock
   c. a dynamic natural body of mineral and organic matter
   d. composed of mineral particles
   e. not a result of weathering processes

2. The basic sampling unit in soil surveys is called a/an

   a. polypedon
   b. pedon
   c. horizon
   d. horizon
   e. profile

3. The R horizon refers to

   a. weathered or unconsolidated bedrock
   b. soils redeveloped through reclamation
   c. a zone of eluviation in the soil horizon
   d. the basic mapping unit of soils in an area
   e. a layer of the soil horizon that is 20 to 30 percent organic material

4. Soil Taxonomy refers to

   a. a method of soil classification based on natural soil processes
   b. a system that has been in use since 1938
   c. a system of classification that we owe to Dukuchaev and the Russians
   d. a 1975 system of soil classification based on observable properties, revised in 1998
   e. the Marbut classification system

5. Which of the following lists is in the proper hierarchy of the Soil Taxonomy system (least to most occurrences)?

   a. orders, suborders, soil series, small groups
   b. soil orders, series, great horizons, small horizons
   c. soil orders, suborders, great groups, subgroups, families, series
   d. pedon, polypedon, horizons, profiles
   e. epipedon, subsurface diagnostic horizon

6. A soil order characteristic of the Amazon Basin is a/an

   a. Alfisol
   b. Aridisol
   c. Spodosol
   d. Oxisol
   e. Amazonosol

7. Salinization is a specific process roughly associated with

   a. Alfisols
   b. Ultisols
   c. Histosol
   d. Aridisols
   e. Mollisols, the soils of central Iowa

8. When limestone is added to Spodosols,

   a. soil pH is lowered to benefit fertility
   b. it firms up a weak soil structure
   c. it adds important colloids to the soil profile
   d. it raises pH, raising crop yields per acre
   e. podzolization occurs

9. The undeveloped soils of Zabriskie Point, Death Valley, are characteristic Entisols.

   a. true
   b. false

10. Gelisols (in the U.S. system) are typical of arctic tundra regions.

    a. true
    b. false

11. Bogs of sphagnum peat are not related to any soil order.

    a. true
    b. false

12. The Kesterson Wildlife Refuge was destroyed by salinization and concentrations of selenium.

    a. true
    b. false

13. Each layer exposed in a pedon is called a soil horizon.

    a. true
    b. false

14. The B horizon normally experiences the greatest concentration of organic matter.

    a. true
    b. false

15. The smallest natural lump of particles is called a ped.

    a. true
    b. false

16. Individual soil aggregates are called soil separates.

    a. true
    b. false

17. About 15,000 soil types have been recognized in the United States alone.

    a. true
    b. false

18. Soil erosion that exceeds 11 metric tons/hectare actually reduces productivity.

    a. true
    b. false

19. Andisols represent a classification of soils derived from volcanic origins.

    a. true
    b. false

20. The Canadian System of Soil Classification (CSSC) uses the same soil orders as the Soil Taxonomy.

    a. true
    b. false

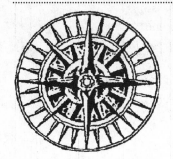

19

ECOSYSTEM ESSENTIALS

The interaction of the atmosphere, hydrosphere, and lithosphere produces conditions within which the biosphere exists. Chapter 19 continues the process of synthesizing all these "spheres" into a complete spatial picture of Earth that culminates in Chapter 20. In this complex age, the spatial tools of the geographic approach are uniquely suited to unravel the web of human impact on Earth's systems. Many career opportunities in planning, GIS analysis, environmental impact assessment, and location analysis are available to those with a degree in geography.

The biosphere extends from the floor of the ocean to a height of about 8 km (5 mi) into the atmosphere. The biosphere is composed of myriad ecosystems, from simple to complex, each operating within general spatial boundaries. Ecology is the study of the relationships between organisms and their environment and among the various ecosystems in the biosphere.

Biogeography, essentially a spatial ecology, is the study of the distribution of plants and animals and the diverse spatial patterns they create across Earth.

A Focus Study details the unique and complex Great Lakes ecosystem. Lake levels are updated through 2004. News Reports cover navigating turtles that can read Earth's magnetic field, Chinstrap penguins' busy summer season, and a new confirmation of biodiversity principles.

OUTLINE HEADINGS AND KEY TERMS

The first-, second-, and third-order headings that divide Chapter 19 serve as an outline for your notes and studies. The key terms and concepts that appear **boldface** in the text are listed here under their appropriate heading in *bold italics*. All these highlighted terms appear in the text glossary. Note

the check-off box (❏) so you can mark your progress as you master each concept. These terms should be in your reading notes or used to prepare note cards. The ✆ icon indicates that there is an accompanying animation on the Student CD.

The outline headings and terms for Chapter 19:

- ❏ *ecosystem*
- ❏ *ecology*
- ❏ *biogeography*

Ecosystem Components and Cycles

✆ **The Global Carbon Cycle**

✆ **The Nitrogen Cycle**

Communities

- ❏ *community*
- ❏ *habitat*
- ❏ *niche*

Plants: The Essential Biotic Component

- ❏ *vascular plants*
- ❏ *stomata*

Photosynthesis and Respiration

- ❏ *photosynthesis*
- ❏ *chlorophyll*
- ❏ *respiration*

Net Primary Productivity

- ❏ *net primary productivity*
- ❏ *biomass*

Abiotic Ecosystem Components

Light, Temperature, Water, and Climate

Life Zones

- ❏ *life zone*

KEY LEARNING CONCEPTS FOR CHAPTER 19

The following key learning concepts help guide your reading and comprehension efforts. The operative word is in *italics*. Use these carefully to guide your reading of the chapter and note that STEP 1 asks you to work with these concepts. These same learning concepts are used in organizing the summary and review section at the end of the chapter—grouping together definitions, a list of key terms, and review questions.

After reading the chapter and using this study guide, you should be able to:

- *Define* ecology, biogeography, and the ecosystem concept.
- *Describe* communities, habitats, and niches.
- *Explain* photosynthesis and respiration, and *derive* net photosynthesis and the world pattern of net primary productivity.
- *List* abiotic ecosystem components, and *relate* those components to ecosystem operations.
- *Explain* trophic relationships in ecosystems.
- *Relate* how biological evolution led to the biodiversity of life on Earth.
- *Define* succession, and *outline* the stages of general ecological succession in both terrestrial and aquatic ecosystems.

✳ STEP 1: Critical Thinking Process

Using your interest and learning, and the following questions as guidelines <u>only</u>, briefly discuss your experience with this chapter. In examining your learning you need not go through each of these questions in detail, simply provide an overview of your critical thinking process as it relates to some aspect of this chapter.

- What did you know about the learning concept before you began?
- Which information sources did you use in your learning (text, class, other)?
- Were you able to complete the action stated in the learning concept? What did you learn?
- Are there any aspects of the concept about which you want to know more?

Critical Thinking and Chapter 19: _____

_____

_____

_____

_____

✳ STEP 2: Ecosystems

1. Define the following:

(a) ecosystem: _____

_____

(b) ecology: _____

_____

(c) biogeography: _____

_____

(d) community: _____

_____

(e) habitat: _____

_____

(f) niche: _____

_____

2. Read the quote that is in the caption to Figure 19.1. The writer beautifully captures the integrated nature of ecosystems and life on Planet Earth. Each decision we make, each move, causes a shudder in the web—life is interconnected and interwoven, from the smallest to the most complex.

3. Examine the variety of plants and animals shown in Figures 19.3 and 19.4 as they work to fit specific niches. The environment is a place of great diversity, all working to survive and flourish, and producing such beauty in form and color.

4. Describe lichen as an example of the principle of *symbiosis*. What does each member of the partnership provide? Is there any aspect of this relationship that might be similar to the biosphere and Earth–human relationships described in Figure 19.5 and text?

_____

_____

_____

5. Complete the labeling for this illustration of abiotic and biotic system components as derived from Figure 19.2a. You may want to augment the illustration with color shading.

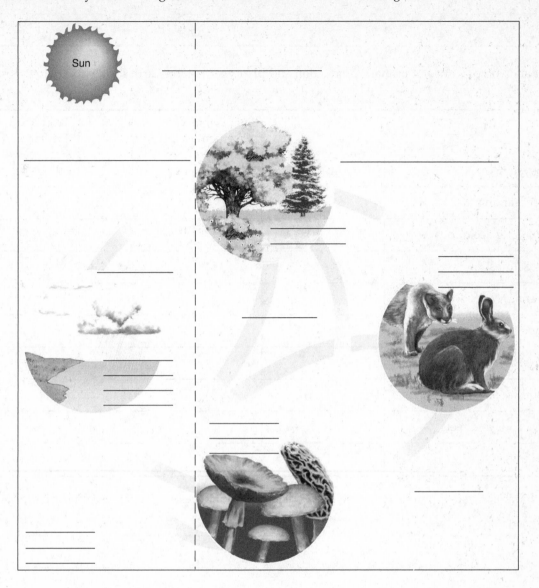

6. In the space provided, record the formulas for *photosynthesis* and *respiration*. Identify each component by symbol and with its name placed below on the next line.

(a) photosynthesis:_____ = _____

_____

(b) respiration:_____ = _____

_____

7. Describe in general terms the distribution of worldwide net primary productivity shown in Figure 19.7 (map and two remote sensing images) and discussed in the text. Relate this distribution to any climatic factors that may assist in explaining the distribution.

_____

_____

_____

_____

8. List and compare the net primary production and plant biomass on Earth for the following ecosystems (Table 19.1):

Ecosystem	Area $(10^6 km^2)$	Net Primary Productivity per Unit Area $(g/m^2/yr)$ Normal Range Mean	World Net Primary Production $(10^9 t/yr)$
Tropical rain forest			
Temperate deciduous			
Boreal forest			
Savanna			
Tundra			
Desert and semidesert			
Cultivated land			
Algal beds and reefs			
Estuaries			

9. As abiotic ecosystem components, briefly describe the importance of each of these physical factors.

(a) light: _____

_____

(b) temperature: _____

_____

(c) water: _____

_____

(d) climate: _____

_____

(e) Earth's magnetic field (News Report 19.1): _____

_____

10. Complete the following diagram showing the relationship between temperature, precipitation, and ecosystems.

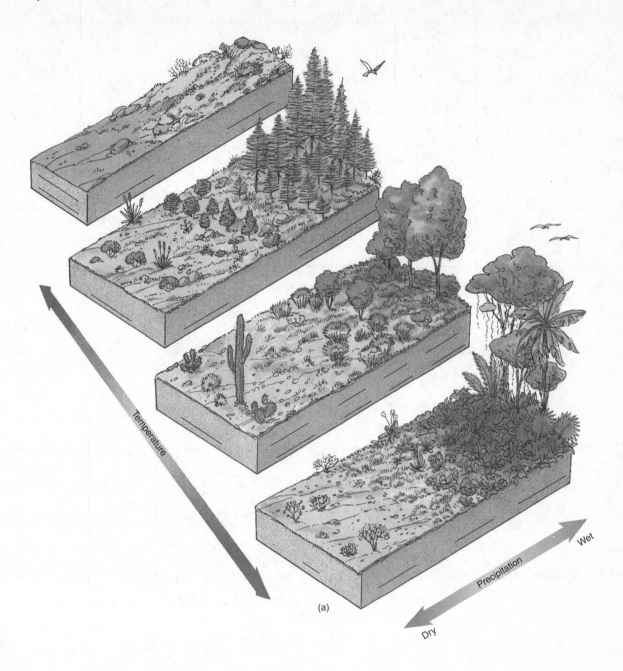

(a)

Temperature

Precipitation

Wet

Dry

11. News Report 19.2 focuses on The Dead Zone. What exactly is the dead zone? What are the biological causes of this phenomenon? What are the cultural causes of this phenomenon? Where does it occur? Is there a seasonal aspect to it? Evaluate the biological and economic benefits of precision agriculture with regards to the dead zone (News Report 1.1). _____

_____

_____

12. Based on the text and Figure 19.10, list a sampling of carbon dioxide sinks (absorption points) and carbon dioxide sources shown in the illustration.

(a) CO_2 sinks: _____

_____

_____

_____

(b) CO_2 sources: _____

_____

_____

_____

13. Based on the text and Figure 19.11, list a sampling of nitrogen sinks (absorption points) and nitrogen sources shown in the illustration.

(a) Nitrogen sinks: _____

_____

_____

_____

(b) Nitrogen sources: _____

_____

_____

_____

14. Using either coast redwoods and maples, or Mallards and the Snail Kite, explain why they have different ranges and distributions based on their *limiting factors* and *tolerance* (Figure 19.12). _____

_____

_____

_____

15. In the space provided, make a <u>simple</u> sketch (in your own style) of the relations among producers, consumers, and decomposers in a typical food chain. Identify each trophic level and activity in your sketch with appropriate labels. Begin by placing primary producers at the lower margin of the box. (Use the text discussion and Figures 19.13, 19.14, and 19.15 for basic information.)

16. Discuss the relationship between Figure 19.16, showing efficiency in food webs, and Figure 19.19, showing the biological magnification of toxins in food chains. _____

_____

_____

_____

✳ STEP 3: Evolution and Ecological Succession

1. In basic terms, describe the process of ecological succession in a community of plants and animals as suggested in the text. What types of disruptions can occur during succession (Figures 19.21, 19.24, and 19.26)?

_____

_____

_____

_____

2. In the 10 photos and satellite image in Figure 19.26 you see comparisons from just before the eruption of Mount St. Helens, and 8 photos comparing the same scene from 1983 to 1999. Take a moment and study these photos. Briefly describe what you see as the differences in the landscapes. How do you assess the rate of recovery?

_____

_____

_____

_____

3. Briefly explain the pioneering work of Dr. David Tilman highlighted in News Report 19.4. What was confirmed? Discovered?

_____

_____

_____

_____

_____

4. Explain the principle of fire ecology and succession using the example of the Yellowstone National Park fire in 1988 (Figures 19.24b, 19.29a, 19.29b, and 19.29d). _____

_____

_____

_____

(a) How have modern fire prevention skills actually worsened the threat of wildfires? _____

_____

_____

5. Explain how agricultural ecosystems differ from natural ecosystems in terms of complexity and ecological succession. _____

_____

_____

_____

6. How far north are climate zones expected to shift in the next 100 years? How could this affect the distribution of plant communities? What would be the results if plant communities cannot migrate at the same pace as climate zones? _____

_____

_____

_____

7. Select one of the Great Lakes discussed in Focus Study 19.1. Briefly detail information about the lake you selected. _____

_____

_____

_____

_____

✳ STEP 4: NetWork—Internet Connection

There are many Internet addresses (URLs) listed in this chapter of the *Geosystems* textbook. Go to any two URLs, or "Destination" links on the *Geosystems* Home Page, and briefly describe what you find.

1. _____: _____

_____

2. _____: _____

_____

SAMPLE SELF-TEST

(Answers appear at the end of the study guide.)

1. A study of the interrelationships between organisms and their environment is called

    a. ecosystems
    b. ecology
    c. community
    d. biogeography

2. A study of the spatial aspects of ecology and the distribution of plants and animals is called

    a. ecosystems
    b. ecology
    c. community
    d. biogeography

3. Which of the following is <u>incorrect</u>?

    a. habitat—specific location of an organism
    b. niche—function of a life-form within a given environment
    c. ecosystem—a self-regulating association of living plants and animals and their nonliving environment
    d. community—specific location of an organism

4. Photosynthesis involves

    a. the release of carbon dioxide in a process using sunlight
    b. carbon dioxide and water under the influence of sunlight
    c. important reactions within the stems and roots of plants
    d. processes that also operate at night

5. The net dry weight of organic matter that feeds a food chain is

    a. not related to net primary productivity
    b. biomass or the useful amount of chemical energy
    c. produced to a greater extent in tundra environments than in savannas
    d. measured in the text in oxygen per square meter per year

6. An example of a biotic ecosystem component is

    a. water
    b. calcium
    c. a geochemical cycle
    d. a heterotroph

7. In a food chain, toxic chemicals tend to accumulate and concentrate in the

    a. top carnivore
    b. autotrophs
    c. primary consumers
    d. primary producers

8. The landscape devastated by the eruption of Mount St. Helens

    a. is recovering through secondary succession
    b. has formed a type of pioneer community, and secondary succession
    c. is now a climax community
    d. will never recover as is evident from an analysis of the land there

9. A mature community is somewhat stable, self-sustaining, and a symbiotically functioning community.

    a. true
    b. false

10. The gradual enrichment of a water body with nutrients and organic materials is called eutrophication.

    a. true
    b. false

11. The final Yellowstone fire report concluded that more fire-fighting equipment and fire suppression is required to prevent further tragedy.

    a. true
    b. false

12. The media and press accurately reported the extent of the fires in Yellowstone.

    a. true
    b. false

13. The interacting population of animals defines only a community.

    a. true
    b. false

14. Gases flow into and out of leaves.

    a. true
    b. false

15. Respiration is opposite in effect to the process of photosynthesis, and is how the plant derives energy for its operations.

    a. true
    b. false

16. Hydrogen, oxygen, and carbon comprise 99% of Earth's biomass.

    a. true
    b. false

17. A physical or chemical component that inhibits biotic activities either through excess or lack is called a limiting factor.

    a. true
    b. false

18. The duration of rainfall and precipitation is known as the photoperiod.

    a. true
    b. false

19. The study of the spatial aspects of ecology and the distribution of plants and animals is called biogeography.

    a. true
    b. false

20. The interacting populations of living plants and animals in an area form a subdivision in nature described as a/an

    a. ecosystem
    b. ecotone
    c. community
    d. niche

21. Chlorophyll _____green light wavelengths.

    a. reflects
    b. absorbs
    c. transmits
    d. uses

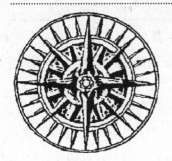

20

TERRESTRIAL BIOMES

Chapter 20 synthesizes many of the elements of *Geosystems*, bringing them together to create a regional portrait of the biosphere. To facilitate this process Table 20.1 is presented to portray aspects of the atmosphere, hydrosphere, and lithosphere that merge to produce the major terrestrial ecosystems. Beyond the specific description of ten major terrestrial biomes this chapter is meant as an overview of the text. Table 20.1 contains columns of vegetation characteristics, soil classes, climate type, annual precipitation range, temperature patterns, and water balance characteristics.

The biosphere is quite resilient and adaptable, whereas many of the specific biomes and communities are greatly threatened by further destructive impacts by society. The irony is that some plants and animals in these biomes possess cures and clues to human disease and mechanisms to recycle the excessive levels of carbon dioxide now entering the atmosphere.

Focus Study 20.1 presents the biodiversity concepts and the biosphere reserve effort to establish protected "islands" of natural biomes. The principal goal is to hold off the record number of extinctions now taking place. Even the most detached individual must recognize that humankind has been an agency of change on Earth—the creators of cultural, albeit artificial, landscapes. Spatial implications of these dynamic trends are of particular importance to physical geographers, for we have the potential for spatial analysis and synthesis. Our discipline is at the heart of geographic information system model construction. The biosphere is quite resilient and adaptable, whereas many specific biomes and communities are greatly threatened by further destructive impacts. The irony is that some plants and animals in these biomes contain cures and clues to human disease, potential new food sources, and

mechanisms to recycle the excessive levels of carbon dioxide now entering the atmosphere.

OUTLINE HEADINGS AND KEY TERMS

The first-, second-, and third-order headings that divide Chapter 20 serve as an outline for your notes and studies. The key terms and concepts that appear **boldface** in the text are listed here under their appropriate heading in ***bold italics***. All these highlighted terms appear in the text glossary. Note the check-off box (❏) so you can mark your progress as you master each concept. These terms should be in your reading notes or used to prepare note cards.

The outline headings and terms for Chapter 20:

Biogeographical Realms

❏ ***biogeographical realm***

Transition Zones

❏ ***ecotone***

Terrestrial Ecosystems

❏ ***terrestrial ecosystem***
❏ ***biome***
❏ ***formation classes***

Earth's Major Terrestrial Biomes

Geographic Scenes: East Greenland Photo Gallery

High Latitude Animals Photo Gallery

Equatorial and Tropical Rain Forest

❏ ***equatorial and tropical rain forest***

URLs listed in Chapter 20

Large Marine Ecosystems (LMEs):
http://www.edc.ure.edu/lme/

National Marine Sanctuaries:
http://www.sanctuaries.nos.noaa.gov/

Exotic species:
http://invasions.bio.utk.edu/bio_invasions/index.html
http://www.dnr.state.mn.us/ecological_services/invasives.html

Tropical Rainforest Coalition:
http://www.rainforest.org/

World Resources Institute forest info:
http://www.wri.org/wri/cat-frst.html

Rainforest Information Center:
http://www.rainforestinfo.org.au/welcome.htm

Rainforest Action Network:
http://www.ran.org/

IUCN:
http://www.redlist.org
http://www.wcmc.org.uk/data/database/rl_anml_combo html
http://www.iucn.org/

Endangered species home page of the Fish and Wildlife Service:
http://endangered.fws.gov/

World Wildlife Fund:
http://www.panda.org/

UNESCO MAB:
http://www.unesco.org/mab/

Man and the Biosphere Species Database Canada and U.S.:
http://www.eman-rese.ca/partners/mab/intro.html
http://www.ice.ucdavis.edu/mab/

MAB Links:
http://www.mabnet.org/misc/links.html

Nature Conservancy:
http://www.nature.org/

KEY LEARNING CONCEPTS FOR CHAPTER 20

The following key learning concepts help guide your reading and comprehension efforts. The operative word is in *italics*. Use these carefully to guide your reading of the chapter and note that STEP 1 asks you to work with these concepts. These same learning concepts are used in organizing the summary and review section at the end of the chapter—grouping together definitions, a list of key terms, and review questions.

After reading the chapter and using this study guide, you should be able to:

- *Define* the concept of biogeographical realms of plants and animals, and *define* ecotone, terrestrial ecosystem, and biome.
- *Define* six formation classes and the life-form designations, and *explain* their relationship to plant communities.
- *Describe* 10 major terrestrial biomes, and *locate* them on a world map.
- *Relate* human impacts, real and potential, to several of the biomes.

✳ STEP 1: Critical Thinking Process

Using your interest and learning, and the following questions as guidelines only, briefly discuss your experience with this chapter. In examining your learning you need not go through each of these questions in detail, simply provide an overview of your critical thinking process as it relates to some aspect of this chapter.

- What did you know about the learning concept before you began?
- Which information sources did you use in your learning (text, class, other)?
- Were you able to complete the action stated in the learning concept? What did you learn?
- Are there any aspects of the concept about which you want to know more?

Critical Thinking and Chapter 20: _____

_____

_____

_____

_____

✳ STEP 2: Biogeographic Realms

1. Explain the significance of biogeographic realms (Figure 20.1) and their relationship to establishing regional patterns of Earth's biomes.

_____

_____

_____

2. Describe the concept and reality of "Large Marine Ecosystems" (News Report 20.1).

_____

_____

_____

3. For review, please *reread* and *recheck* your notes for the definitions of the following terms in Chapter 19: *ecosystem*, *ecology*, *biogeography*, *community*, *habitat*, and *niche*. Then, define the following:

(a) biome: _____

_____

(b) formation classes: _____

_____

✳ STEP 3: Terrestrial Ecosystems

1. Given the description of the equatorial and tropical rain forest in the text, examine the art and photographs in Figures 20.4 and 20.5. Can you identify any of the features described by the text in these photographs? Explain.

_____

_____

_____

2. Present a brief overview of the condition and status of deforestation in the tropics (Figure 20.6 and Focus Study 20.1). Why is the diversity of the rain forest biome so critical to humanity?

_____

_____

_____

_____

3. From the comprehensive Table 20.1, the world map in Figure 20.3, and following a careful reading of the chapter, determine the biome that best characterizes each of the following descriptions and record its name in the space provided. This exercise provides an overview of the text and integrates many of the elements of physical geography that we have studied together.

An annual precipitation range of 25–75 cm: _____

POTET greater than 1/2 PRECIP: _____

Southern and eastern evergreen pines: _____

Mollisols and Aridisols: _____

Temperate with a cold season: _____

Characteristic of central Australia: _____

Selva: _____

Sclerophyllous shrub: _____

Short summer, cold winter: _____

Characteristic of the majority of central Canada: _____

Transitional between rain forest and tropical steppes: _____

Tallest trees on Earth: _____

Sedges, mosses, and lichens: _____

Characteristic of Zambia (south central Africa): _____

Four biome types that occur in Chile:_____

Less than 40 rainy days in summer: _____

Characteristic of Tennessee: _____

Precipitation of 150–500 cm/year, outside the tropics: _____

Always warm, water surpluses all year: _____

Characteristic of central Greenland: _____

Characteristic of Iran (northeast of the Persian Gulf): _____

Characteristic of northern Mexico: _____

Major area of commercial grain farming: _____

Seasonal precipitation of 90 to 150 cm/year: _____

Characteristic of Quebec, Canada: _____

Characteristic of southern Argentina: _____

Spodosols and permafrost, short summers: _____

Southern Spain, Italy, and Greece: _____

Characteristic of Ireland and Wales: _____

Uruguay and the La Plata: _____

Just west of the 98th meridian in the U.S.: _____

Just east of the 98th meridian in the U.S.: _____

Bare ground and xerophytic plants: _____

East coast of Madagascar: _____

West coast of Madagascar: _____

The bulk of Cuba: _____

Deciduous needleleaf trees: _____

The majority of Indonesia: _____

4. Characterize the present black rhino and white rhino populations of Africa. Include in your response the concept of *biodiversity* and the potential role that *biosphere reserves* might play in the future of these species (Focus Study 20.1).

_____

_____

_____

5. What surprised you about the information in Table 1 in Focus Study 20.1? Explain.

_____

_____

_____

6. News Report 20.2 is about "Alien Invaders of Exotic Species." Is this a low-budget sci-fi movie or a serious ecological problem? Briefly explain the contents of this news report. Make a few suggestions about possible solutions. _____

_____

_____

_____

7. What would be the environmental effects of drilling for oil in the ANWR? Compare the projected oil supply to current U.S. demand and to the amount that could be saved by increasing automobile mileage efficiency. _____

_____

_____

_____

_____

✳ STEP 4: NetWork—Internet Connection

There are many Internet addresses (URLs) listed in this chapter of the *Geosystems* textbook. Go to any two URLs, or "Destination" links on the *Geosystems* Home Page, and briefly describe what you find.

1. _____ : _____

_____

2. _____ : _____

_____

SAMPLE SELF-TEST
(Answers appear at the end of the study guide.)

1. The regions where groups of species evolved and from which they diffuse and migrate are called

    a. ecosystems
    b. Wallace limits
    c. biogeographical realms
    d. communities
    e. biomes

2. The active layer in the sea that is analogous to the canopy of the rain forest is called a/an

    a. photic layer
    b. aquatic ecosystem
    c. large marine ecosystem
    d. there was no analogy drawn in the text
    e. life zone

3. Which of the following is <u>incorrect</u>?

    a. community—interacting populations of plants and animals
    b. ecosystems—interplay between communities and the physical environment
    c. life-form—outward physical properties of plants
    d. formation class—characterized by a dominant plant community
    e. LME—an early life-form in the tropics

4. A biome is

    a. an ecosystem characterized by related animal populations
    b. a large, stable terrestrial ecosystem or aquatic ecosystem
    c. the smallest local designation of a community
    d. a natural community, most of which are unaffected by human activity
    e. mapped as defined by abiotic components such as climate

5. Relative to deforestation of the rain forests,

    a. activities have not slowed, and annually exceed 169,000 km^2(65,000 mi^2)
    b. it has markedly slowed from previous highs due to international efforts
    c. it occurs principally for lumber and abundant specific tree species
    d. modern cutting rather than burning methods are in use
    e. one-tenth of the forests are now gone, including 10 percent of Africa's

6. The biosphere reserve effort specifically involves principles of

    a. geography
    b. biology
    c. ecology
    d. geology
    e. island biogeography

7. The pines of the southeastern United States are characteristic of which biome?

   a. tropical savanna
   b. Mediterranean shrubland
   c. temperate rain forest
   d. midlatitude broadleaf and mixed forest
   e. midlatitude grasslands (tall grass)

8. Desertification is thought to be principally due to

   a. poor agricultural practices and overgrazing
   b. salinization
   c. global climate change
   d. shifts in Earth's orbit
   e. political conflicts

9. The creosote bush is thought to control its own watershed with toxins.

   a. true
   b. false

10. The tundra is found both in the extreme latitudes of North America and in Russia.

    a. true
    b. false

11. George Perkins Marsh published the conservation epic *The Earth as Modified by Humans* in 1874.

    a. true
    b. false

12. There are presently 30 large marine ecosystems designated, with more than 200,000 km^2 so identified.

    a. true
    b. false

13. A terrestrial ecosystem is a known as a biome.

    a. true
    b. false

14. The forests in the Amazon are also called the selva.

    a. true
    b. false

15. The floor of the equatorial rain forest is usually thick with smaller trees.

    a. true
    b. false

16. Caatinga, brigalow, and dornveld are types of chaparral associated with Mediterranean lands.

    a. true
    b. false

17. The principal division line between the tall grass and short grass prairies is the 51 cm (20 in.) annual precipitation isohyet.

    a. true
    b. false

18. Ukraine is an example of a midlatitude grassland.

    a. true
    b. false

21

EARTH AND THE HUMAN DENOMINATOR

We come to the end of our journey through the pages of *Geosystems* and an introduction to physical geography. A final capstone chapter is appropriate given the dynamic trends that are occurring in Earth's physical systems. Our vantage point in this course of study is physical geography. We examine Earth through its energy, atmosphere, water, weather, climate, endogenic and exogenic systems, soils, ecosystems, and biomes, all of which leads to an examination of the most abundant large animal: *Homo sapiens*.

Chapter 21 is newly titled to reflect the current assessment being made by science of planetary impacts by modern societies. Because human influence is pervasive, we consider the totality of our impact the *human denominator*. Just as the denominator in a fraction tells how many parts a whole is divided into, so the growing human population and the increasing demand for resources and rising planetary impact suggest how much the whole Earth system must adjust. Yet Earth's resource base remains relatively fixed.

During the first edition of *Geosystems* the largest-ever gathering of nations and individuals took place—the Earth Summit of 1992. During the fifth edition a second Earth Summit was held. Many aspects of physical geography were at the heart of this meeting and the subject of the agreements reached.

We hope this last chapter gives you something to contemplate as you leave this course. This is a time of great change in natural physical systems and therefore a time that demands a geographic perspective.

OUTLINE HEADINGS AND KEY TERMS

The first-order headings that divide Chapter 21 serve as an outline for your notes and studies.

The outline headings for Chapter 21:

The Human Count and the Future

An Oily Bird

The Need for International Cooperation

Twelve Paradigms for the 21st Century

Who Speaks for Earth?

News Report and High Latitude Connection

News Report 21.1: Gaia Hypothesis Triggers Debate

High Latitude Connection 21.1: Report from Reykjavik—Arctic Climate Impact Assessment

---

URLs listed in Chapter 21

U.S. Bureau of Census *PopClock Projection:*
http://www.census.gov/cgi-bin/popclock

Population Reference Bureau:
http://www.prb.org

Earth Summit:
http://www.earthsummit2002.org/

Arctic Climate Impact Assessment:
http://www.acia.uaf.edu

Arctic Council:
http://www.arcticcouncil.org

Arctic Monitoring and Assessment Program:
http://www.amp.no

Conservation of Flora and Fauna:
http://www.caff.is

International Arctic Science Committee:
http://www.iasc.no

KEY LEARNING CONCEPTS FOR CHAPTER 21

The following key learning concepts help guide your reading and comprehension efforts. The operative word is in *italics*. Use these carefully to guide your reading of the chapter and note that STEP 1 asks you to work with these concepts. These same learning concepts are used in organizing the summary and review section at the end of the chapter—grouping together definitions, a list of key terms, and review questions.

After reading the chapter and using this study guide, you should be able to:

- *Determine* an answer for Carl Sagan's question, "Who speaks for Earth?"
- *Describe* the growth in human population, and *speculate* on possible future trends.
- *Analyze* "An Oily Bird," and *relate* your analysis to energy consumption patterns in the United States and Canada.
- *List* the subjects of recent environmental agreements, conventions, and protocols, and *relate* them to physical geography and Earth systems science (geosystems).
- *List* 12 paradigms for the 21st century.
- *Appraise* your place in the biosphere, and *realize* your relation to Earth systems.

❋ STEP 1: Critical Thinking Process

Using your interest and learning, and the following questions as guidelines <u>only</u>, briefly discuss your experience with this chapter. In examining your learning you need not go through each of these questions in detail, simply provide an overview of your critical thinking process as it relates to some aspect of this chapter.

- What did you know about the learning concept before you began?
- Which information sources did you use in your learning (text, class, other)?
- Were you able to complete the action stated in the learning concept? What did you learn?
- Are there any aspects of the concept about which you want to know more?

Critical Thinking and Chapter 21: _____

_____

_____

_____

_____

❋ STEP 2: Human Population

1. Briefly characterize the growth of the human population in terms of the interval in years to achieve each new billion in the count (Figure 21.4). What does the future hold in terms of continued growth?

_____

_____

_____

_____

2. From Table 21.1, compare and contrast the population doubling time for the world, MDCs and LDCs (with and without China). Do you think that these growth rates should be increased, reduced, or maintained? How do you think this should best be done?

_____

_____

_____

_____

3. According to the Internet URL given in the text (*http://www.census.gov/cgi-bin/popclock*), what is the

present human population of Earth? _____

_____

Of the United States? _____; of Canada? _____ Where

is the Canadian data available? _____

4. Compare the projected population growth in more developed (MDCs) and less developed countries (LDCs) in the next 50 years. Compare the environmental effects of the populations of MDCs and LDCs using

the $I=P \cdot A \cdot T$ concept. _____

_____

_____

_____

_____

✳ STEP 3: An Oily Bird

1. Beginning with the oil-contaminated Western Grebe shown in Figure 21.6, outline the chain of events, linkages, and overall system between the sick bird, a tanker, imports, a gas station, your own local neighborhood shopping mall, and transportation patterns.

_____

_____

_____

_____

[Interesting to note: one scientist calculated that if the U.S. automobile and light truck (SUV) fleet increased 2.7 miles per gallon in efficiency—or to about 30 mpg average—it would be the equivalent to all the oil that is now imported from the Middle East.]

2. Overview the death toll of animals caused by the *Exxon Valdez* spill (Figure 21.6).

_____

_____

3. According to the text, briefly characterize remaining oil reserves.

_____

_____

4. Given items 1, 2, and 3 just completed, and knowing the other impacts of fossil fuel combustion on the atmosphere, climate, precipitation chemistry, and land, briefly state *your point of view* as to what society should do either to alleviate these problems or reduce their impacts, or to continue business as usual. Be as specific as possible about your suggested policies on the issues. There is no right or wrong answer here, just your answer!

_____

_____

_____

_____

_____

✳ STEP 4: The Need for International Cooperation

During June 1992, in Rio de Janeiro, a first-ever global conference called the *United Nations Conference on Environment and Development (UNCED)* took place. The purpose was to address with substance the many fundamental issues that relate to achieving sustainable world development. The historic conference and a summary of the five principal accomplishments are reviewed in the chapter. A decade passed before the world chose to meet again in 2002 to discuss the fate of Earth and human society. The URL for the Earth Summit 2002, held in Johannesburg, South Africa is **http://www.earthsummit2002.org**. The agenda included climate change, freshwater, gender issues, global public goods, HIV/AIDS, sustainable finance, and the five Rio Conventions.

1. Go to the URL mentioned above and report on the outcome of this second Earth Summit. What were the main accomplishments, failures, and new agenda items that emerged? What is your opinion about such meetings? Do you think that there should be a third Earth Summit?

_____

_____

_____

_____

(a) From the Web site, or from other sources, characterize the performance and participation of the United States in the summit; then characterize the Canadian role. Assess positive or negative opinions you encounter.

_____

_____

_____

_____

2. Who were the parties involved with the Arctic Climate Impact Assessment (ACIA)? Which countries were involved? How long did it take for the report to be produced? How many scientists contributed to the report?

_____

_____

3. How did the findings of the ACIA compare with the findings of the IPCC? Which model scenario did the ACIA use? How do the effects of global warming in the Arctic compare with the effects of global warming in non-arctic regions?

_____

_____

_____

_____

4. Please take a moment to read and assess the list in the chapter of *Twelve Paradigms for the 21st Century*. The items are not ranked in any particular order. In your opinion, which items are most significant?

_____

_____

(a) Do you think the list omitted something you deem vital? If so, briefly describe it:

_____

_____

_____

A FINAL THOUGHT

Chapter 21 demonstrates an increasing awareness of Earth and environmental issues. More than a decade ago, in 1989, *Time* magazine named Earth the "Planet of the Year" instead of its normal practice of naming a prominent citizen as its person of the year (Figure 21.9). Earth Day, every April 22, is celebrated by more and more people across the globe. Local communities, governmental institutions, and corporations are overwhelmed by the willingness of the public to reduce, reuse, and recycle resources. Governments and vested interests consistently lag behind, or outright oppose the public's growing concern for the environment and increasing desire for sustainable behavior and long-term perspectives. In 2002, Honda and Toyota, without government support, are finding they can't manufacture enough 50+ mpg hybrid cars to keep up with demand!

United Nations Secretary General Kofi Annan, speaking to the Association of American Geographers annual meeting, March 1, 2001, offered this assessment,

> As you know only too well the signs of severe environmental distress are all around us. Unsustainable practices are woven deeply into the fabric of modern life. Land degradation threatens food security. Forest destruction threatens biodiversity. Water pollution threatens public health, and fierce competition for freshwater may well become a source of conflict and wars in the future. ... the overwhelming majority of scientific experts have concluded that climate change is occurring, that humans are contributing, and that we cannot wait any longer to take action ... environmental problems build up over time, and take an equally long time to remedy.

We again quote from this same talk in Chapter 21 when he stated,

> The idea of interdependence is old hat to geographers, but for most people it is a new garment they are only now trying on for size. Getting it to fit—and getting it imprinted on the mental maps that guide our voices and our choices—is one of the crucial projects of human geography for the 21st century. I look forward to working with you in that all-important journey.

Earth-system scientists are assessing the environment with increasing sophistication. Many of these techniques are mentioned throughout the text. Despite this scientific activity, measurement, and confirmation, there remain a few media personalities who have for unexplained reasons captured popular interest by declaring that none of the changes in the ozone layer, or increases in acid deposition, or the dynamics of climate change, among other issues, are actually occurring.

The anti-science "pop" rhetoric became so intense that the American Association for the Advancement of Science (AAAS) ran a strong condemnation of the media circus. Please see: Gary Taubes, "The Ozone Backlash," *Science*, 260 June 11, 1993: 1580–83. The article details the misrepresentation and misunderstanding by the pop critics as they formulated their ill-founded and loudly stated views that everything is just swell. Science and scientists are portrayed in a negative way in many movies. Add to this a general lack of geographic knowledge, and you get an idea of the problem we face in education!

The challenge to all of us, especially those in the leading economies that extract most of Earth's resources, is to learn, apply, and behave in a more responsible manner than has dominated the industrial revolution to date, in ways that consider the future.

Carl Sagan answered his question that opened the chapter, "Who speaks for Earth?" "We speak for Earth," he answered. We are the Earthlings. May we all perceive our spatial importance within Earth's ecosystems and do our part to maintain a life-supporting Earth into the future.

Please feel free to communicate questions, ideas, opinions, and your thoughts about the subjects we have shared in the sixth edition of *Geosystems, An Introduction to Physical Geography*. We will attempt to respond, revise, correct, and update future editions of the text and this Study Guide and look forward to your questions and comments. You can reach us at the e-mail and mailing addresses listed below. You can also interact through our Home Page on the World Wide Web.

The best to you, your academic career, and your future—fellow Earthling. *Carpe diem!*

OUR ADDRESS INFORMATION

Robert W. Christopherson
P.O. Box 128
Lincoln, CA
E-mail: bobobbe@aol.com
http://www.prenhall.com/christopherson

Charles E. Thomsen
American River College
4700 College Oak Drive
Sacramento, CA 95841
E-mail: thomsec@arc.losrios.edu
http://ic.arc.losrios.edu/~thomsec

ANSWER KEY TO SELF-TESTS

Chapter 1

1. c
2. a
3. b
4. d
5. d
6. b
7. c
8. d
9. b
10. d
11. a
12. b
13. a
14. a
15. b
16. a
17. b
18. spatial
19. open; closed; closed
20. Polaris (North Star); Southern Cross (Crux Australis)
21. Primary standard; *NIST-F1*, National Institute for Standards and Technology
22. (Use definitions from glossary for great circle, small circle)
23. a) cylindrical, b) conic, c) planar, d) oval
24. a
25. written scale; representative fraction; graphic (bar) scale
26. matter; energy
27. **(a)** Robinson, oval
    (b) Millers, cylindrical
    (c) Gnomonic, planar
    (d) Modified Goode's homolosine, oval

Chapter 2

1. b
2. d
3. b
4. d
5. d
6. b
7. d
8. c
9. a
10. a
11. a
12. d
13. a
14. b
15. a
16. gamma, x-rays, ultraviolet, visible light, and infrared

Chapter 3

1. b
2. a
3. b
4. a
5. b
6. b
7. c
8. d
9. d
10. c
11. b
12. d
13. c
14. c
15. b
16. b
17. b
18. a
19. b
20. b
21. b

Chapter 4

1. b
2. c
3. d
4. c
5. e
6. c
7. c
8. a
9. b
10. e
11. e

12. 80–95% for fresh snow; 5–10% for asphalt; and 6–8% for the Moon

13. CO_2, H_2O vapor, CH_4, N_2O, CFCs

14. microclimatology, including boundary layers climates

15. a
16. b
17. b
18. a
19. a
20. b
21. b
22. a

Chapter 5

1. d
2. c
3. c
4. c
5. d
6. d
7. d
8. b
9. a
10. a
11. b
12. b
13. a
14. a
15. a
16. −273°C, −459.4°F, 0 K; 32°F (0°C, 273 K); 212°F (100°C, 373 K); United States.
17. Vostok, Antarctica, 78°S 106°E, −89°C (−129°F), July 21, 1983.
18. Al'Aziziyah, Libya, 32°N 13°E, 58°C (136°F), September 13, 1922.
19. and 20. Personal response about local conditions and weather station.

Chapter 6

1. a
2. b
3. b
4. a
5. c
6. c
7. c
8. a
9. c

10. b
11. a
12. b
13. a
14. b
15. b
16. b
17. a
18. b
19. a
20. b
21. a
22. The piling up of ocean water along the western margin of each ocean basin produced by the trade winds that drive the oceans westward.
23. clockwise circulation, the Pacific gyre
24. pressure gradient; Coriolis; friction; gravitational

Chapter 7

1. c
2. c
3. c
4. d
5. c
6. d
7. a
8. d
9. c
10. d
11. c
12. d
13. b
14. d
15. c
16. d
17. b
18. b
19. a
20. a
21. b
22. b
23. a
24. a
25. b
26. hair hygrometer; sling psychrometer; vapor pressure; specific humidity.
27. b
28. c

Chapter 8

1. Personal response related to present location.
2. a
3. a
4. e
5. a
6. d
7. a
8. b
9. a
10. a
11. b
12. a
13. b
14. a
15. b
16. a
17. windward; leeward
18. m; c
19. 19; 11; 2nd
20. Felix, Luis, and Opal
21. b
22. a
23. d
24. d
25. b

Chapter 9

1. b
2. b
3. c
4. d
5. d
6. c
7. c
8. a
9. d
10. d
11. d
12. c
13. b
14. a
15. b
16. a
17. a
18. a

Chapter 10

1. d
2. c
3. d
4. d
5. d
6. c
7. d
8. a
9. d
10. a
11. d
12. b
13. c
14. a
15. c
16. a
17. a
18. b
19. b
20. a
21. a
22. a
23. b
24. b
25. Personal response related to present location.
26. a
27. b

Chapter 11

1. c
2. d
3. b
4. d
5. b
6. d
7. a
8. d
9. b
10. a
11. b
12. b
13. a
14. b
15. a
16. a

17. Uniformitarianism; present ... past; catastrophism.
18. mineral; 99%; rock.
19. b
20. a
21. d
22. a
23. a
24. b

Chapter 12

1. a
2. d
3. c
4. d
5. c
6. b
7. a
8. a
9. a
10. c
11. b
12. a
13. a
14. b
15. b
16. a
17. d
18. c
19. d
20. a

Chapter 13

1. c
2. d
3. e
4. a
5. d
6. a
7. a
8. c
9. a
10. a
11. a
12. a
13. a
14. a
15. a

16. a
17. geomorphology
18. Karst topography; Krs
19. soil creep
20. scarification
21. c
22. b
23. c
24. c

Chapter 14

1. b
2. a
3. b
4. a
5. c
6. b
7. b
8. b
9. a
10. a
11. b
12. b
13. a
14. b
15. a
16. a
17. b
18. a
19. Atlantic; Gulf of Mexico; Pacific; Pacific-Bering Sea
20. parallel; rectangular; radial; annular
21. Atchafalaya River
22. 7 forms; 500 years
23. b
24. a
25. b
26. d
27. a

Chapter 15

1. c
2. c
3. a
4. c
5. c
6. b
7. d

8. d
9. b
10. b
11. a
12. b
13. a
14. a
15. a
16. a
17. b
18. a

Chapter 16

1. d
2. c
3. c
4. a
5. b
6. d
7. d
8. c
9. a
10. b
11. a
12. a
13. b
14. b
15. a
16. b
17. b
18. b
19. (See labels in Fig. 16.1.1, Focus Study 16.1)
20. transition; translation
21. spit; bay barrier; tombolo
22. salt marshes; mangrove swamps

Chapter 17

1. c
2. a
3. e
4. b
5. c
6. d
7. d
8. d
9. a
10. a
11. a

12. b
13. b
14. b
15. a
16. a
17. b
18. a
19. a
20. arete; col; horn; tarn; paternoster lakes.

Chapter 18

1. c
2. b
3. a
4. d
5. c
6. d
7. d
8. d
9. a
10. a
11. b
12. a
13. a
14. b
15. a
16. b
17. a
18. a
19. a
20. b

Chapter 19

1. b
2. d
3. d
4. b
5. b
6. d
7. a
8. b
9. a
10. a
11. b
12. b
13. b
14. a
15. a

16. a	6. e
17. a	7. d
18. b	8. a
19. a	9. a
20. c	10. a
21. a	11. a
	12. a
Chapter 20	13. a
	14. a
1. c	15. b
2. a	16. b
3. e	17. a
4. b	18. a
5. a